D0816989

CISM COURSES AND LECTURES

Series Editors:

The Rectors of CISM
Sandor Kaliszky - Budapest
Mahir Sayir - Zurich
Wilhelm Schneider - Wien

The Secretary General of CISM
Giovanni Bianchi - Milan

Executive Editor
Carlo Tasso - Udine

The series presents lecture notes, monographs, edited works and proceedings in the field of Mechanics, Engineering, Computer Science and Applied Mathematics.
Purpose of the series is to make known in the international scientific and technical community results obtained in some of the activities organized by CISM, the International Centre for Mechanical Sciences.

INTERNATIONAL CENTRE FOR MECHANICAL SCIENCES

COURSES AND LECTURES - No. 389

COMPONENT RELIABILITY UNDER CREEP-FATIGUE CONDITIONS

EDITED BY

JÁNOS GINSZTLER
TECHNICAL UNIVERSITY OF BUDAPEST

R. PETER SKELTON
IMPERIAL COLLEGE LONDON

SpringerWienNewYork

The expenses for the production of this volume have been partially
covered with the financial support of the European Commission
(contract n. ERBFMMACT970206).

Le spese di stampa di questo volume sono in parte coperte da
contributi del Consiglio Nazionale delle Ricerche.

This volume contains 198 illustrations

In order to make this volume available as economically and as
rapidly as possible the authors' typescripts have been
reproduced in their original forms. This method unfortunately
has its typographical limitations but it is hoped that they in no
way distract the reader.

ISBN 3-211-82914-8 Springer-Verlag Wien New York

PREFACE

Materials in modern power generating, processing plant and aero-engines are frequently run at very high temperatures and can undergo severe transient stresses during start-up and shutdown operations. Not only can this lead to unwanted fatigue cracking at the component surface but it is also known that long periods of steady running under load can give rise to internal damage, cavitation and cracking. Failure prevention, residual life assessment and life extension of materials in components operating at high temperatures are becoming increasingly important problems in power plant and associated industries.

Lifetime assessment procedures involve both limits to material deformation and degree of cracking and can be applied:

- *at the design stage*
- *to extend the lifetime of components when the original endurance has been achieved*
- *retrospectively in order to explain how a given component has failed.*

Earlier design and assessment rules were based on empirical formulations but nowadays assessment is based more securely on behaviour (cracking, deformation, cavitation, strength) explained from a material science point of view.

The aim of the Summer School held at CISM, Udine between June 30 and July 4, 1997 was to provide a state-of-the-art exposition on the problems of crack initiation and growth at high temperature and to present experimental and theoretical results which can form the basis of further research, and which can also be applied in practice in order to prevent accumulation of internal damage and failure in component materials. During the course, industrial examples were introduced to illustrate the application of the subjects covered. It thus becomes clear that satisfactory progress can only be made in high temperature lifetime prediction by a combination of:

- *a proper understanding of the principles of time-dependent and time-independent deformation and the fundamentals of fracture*
- *good records of the component history in service, e.g. temperature changes and running schedules*
- *key high temperature tests in the laboratory which are undertaken either to supply the characterising materials data used in the assessments or to support lifetime prediction calculations*
- *the development of simple finite element or other computer routines, as appropriate.*

The course was addressed to professionals active in the design and operation of power plants and those involved in supporting research and development activities in high temperature materials. It was arranged by experts on those aspects of fracture mechanics which are applicable to high temperature problems, experimentalists in high temperature fatigue and thermal shock, advisors in the choice of materials and long term condition monitoring of alloys in plant and theoreticians in the fundamental properties of materials.

The plan of the book comprising the lectures delivered is as follows. After briefly dealing with material selection and material parameters, Czoboly moves on to consider the basic contributions to deformation, namely the elastic and plastic components and the roles of dislocation glide, twinning, stress-directed diffusion, dislocation climb and grain boundary sliding at elevated temperatures. Further sections deal with material testing and theories of fracture, especially the absorbed specific energy. The fundamental differences between fatigue and creep tests are explored.

In a similar manner, the chapter by Nikbin treats of the development of fracture mechanics principles to predict creep crack growth in components i.e. where there is the possibility that defects may initiate and grow in service. In this way, lifetime of cracked and uncracked high temperature components may be calculated and residual life assessments may be made using the principles of damage accumulation. Amongst the topics considered are the stress analysis of cracked bodies, elastic-plastic fracture mechanics, creep fracture mechanics, the laboratory measurement of creep crack growth, models for creep crack growth, stress re-distribution effects on creep crack growth and finally the effect of fatigue on creep crack growth.

The equally important subject of the initiation and growth of fatigue cracks by sudden temperature transients e.g. thermal shock events is next treated by Skelton, following the course of a defect from its initiation and growth in the 'short crack' regime to its eventual complete passing through a body. The order of approach is an introduction to thermal fatigue followed by stress analysis and supporting laboratory data, the prediction of fatigue crack initiation, and subsequent growth into the 'deep crack' regime. A large section is devoted to the case study of a component which failed in service. As elsewhere in the book, attention is paid to the linear damage theories of creep-fatigue and the concept of accumulating internal damage in order to assist in lifetime prediction.

In the contribution by Ainsworth, great use is made of the reference stress concept as an aid to the behaviour of creep crack growth in components of differing geometries, and the application of the now familiar C* parameter which can be obtained by either reference stress or numerical methods. Creep crack initiation time is also estimated by a development of this parameter. Again, the accumulation of internal damage is applied to the case of creep crack growth and an outline is given of the various engineering procedures available for undertaking assessments. These principles are finally applied by working through several examples, culminating in a full case study on creep crack growth.

The final chapter by Ginsztler addresses typical operating conditions of pressure vessels and pipe-line elements, structural problems associated with power plant materials, the accumulation of internal damage and the possible rejuvenation of components by heat treatment. Creep-fatigue interactions (i.e. transgranular and intergranular cracking) are described, as is the process of thermal strain ageing and crack initiation during low cycle thermal shock fatigue. The section ends with a description of failure prevention methods and the possibilities of life extension of plant by undertaking revalidation tests.

Common to all the chapters is the underlying belief that the build-up of internal damage in alloys of interest takes place whether cracks are present or not and the quantification of this process goes a long way towards the satisfactory prediction of component behaviour in service.

The editors would like to thank the Organising Committee at CISM, especially Professor S. Kalisky, for making the stay at Udine both for lecturers and attendees a memorable occasion.

János Ginsztler
R. Peter Skelton

CONTENTS

Page

LIFETIME ASSESSMENT AT HIGH TEMPERATURES

J. Ginsztler
Technical University of Budapest, Budapest, Hungary

Abstract

A classical power plant was generally expected to last for about 20 years or more (nuclear plant for 35-40 years). It would not normally be in operation continuously throughout the whole of this time, however, and the design was usually based on a life of 100000 hours, which is about 11 years.

Typical operating conditions of the different power plant units have changed during the last decades, they often differed from the conditions assumed at the design state. This fact was one of the reasons, why the damage analysis, failure prevention methods and life extension possibilities became so important in the nineties.

Role of operating conditions of pressure vessels, pipeline elements on residual life of power-plant components.

Combustion processes in furnace installations are characterised by parameters varying with time, thus fluctuating more or less around the required value. This concerns velocity of the air flowing into the combustion chamber as well as the fuel, thus fuel-air concentration will change in space and time resulting in pressure and temperature fluctuations.
The characteristics of different operational types in power plants are shown in Fig.1.

The pulsation of burning in pulverised coal fuelled equipment indicates somewhat periodical burning disturbance with frequency independent of equipment geometry, depending mainly on mixture formation, gasification and ignition processes. Therefore, anomalies of this type may be explained from the burning and flow, with fluctuation frequency mainly depending on fuel quality.

Pressure fluctuation coming with internal furnace instability is dangerous from the aspect of safety, because it soon results in repeated stress failure in structural materials of the boiler. Therefore, it is essential to know the elasticity and eigenfrequency of boiler construction units, especially the membrane wall of the furrnace.

According to their origin, the stresses can be divided into mechanical and thermal stresses. Mechanical stresses are caused by internal pressure, by bending moments and by external forces. In terms of mechanical stress of the boiler's structural elements it is essential to observe continuously parameters influencing formation and variation of internal furnace instability.

Based on these registration results, one can set out to build up the general dynamical and identification model of various boiler types. According to models generally built up for furnace systems one can elaborate in detail the model of input / output processes necessary for statistical characterisation of mechanical stress in the boiler. Furthermore, we can analyse stress processes in the combustion chamber, and predict the formation and variation of stress processes for each structural element.

It is worthy to collect data necessary for the formation of a mathematically detailed input / output model based on the general identification model of furnace processes, which give the range of parameters characterising coal quality for several years. Using the models and having the database, we can perform simulations which give the possible range of pressure fluctuations in the combustion chamber.

In this way we can predict the mechanical stress values for individual boiler types and coal qualities. Furthermore, the probability of accidental breakdown of power plant blocks can be decreased and a more economic and safe operation accomplished. (1)

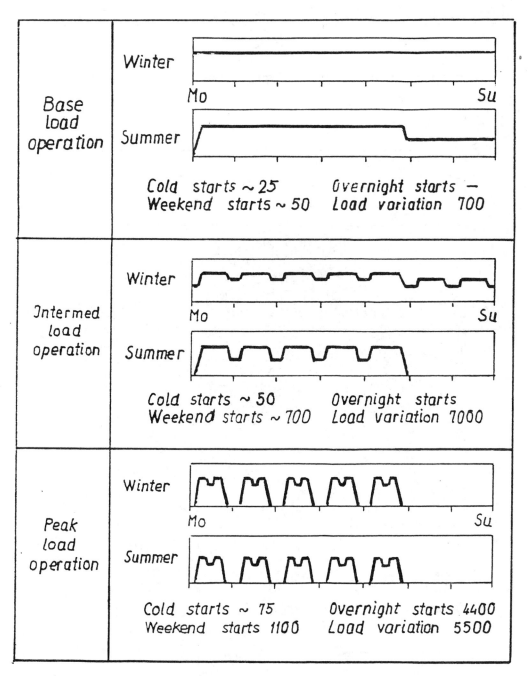

Fig. 1.

Structure of load curve in summer in Hungary in 1992

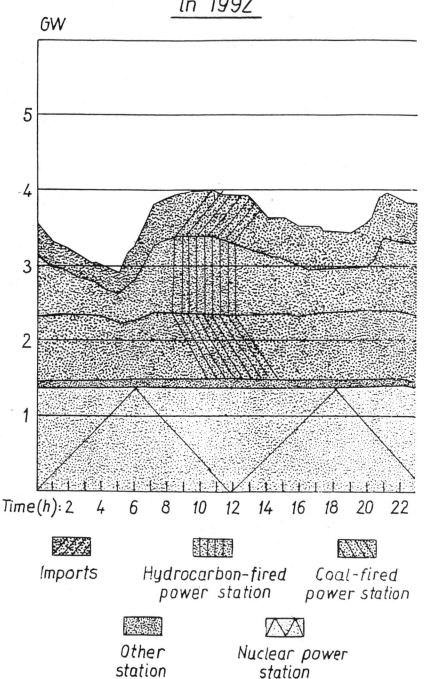

GW

Time(h): 2   4   6   8   10   12   14   16   18   20   22

Imports

Hydrocarbon-fired power station

Coal-fired power station

Other station

Nuclear power station

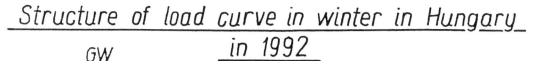

Structure of load curve in winter in Hungary in 1992

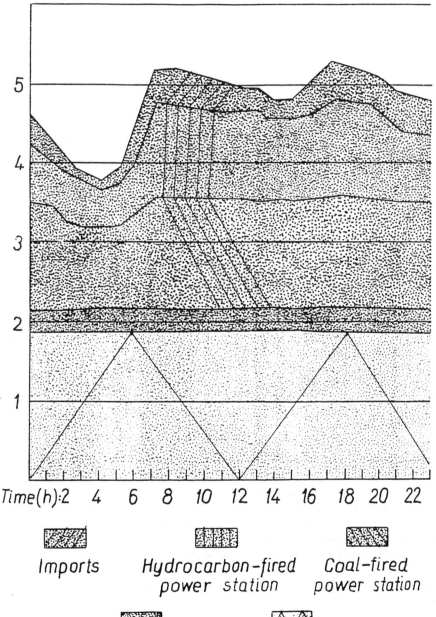

GW

Time(h):2 4 6 8 10 12 14 16 18 20 22

 Imports

 Hydrocarbon-fired power station

Coal-fired power station

Other station

Nuclear power station

Structural problems, associated with power plant materials.

Creep mechanisms, creep-fatigue interactions.

Creep deformation of metals occurs mainly in three ways: firstly by slip involving the movement of dislocations on certain crystallographic planes; secondly, by the formation of small slightly misoriented sub-grains within the original grains of the metal as a result of the climbing of dislocations out of their slip planes to form small angle boundaries; and thirdly by grain boundary sliding.

At lower temperatures slip is the predominant mode of deformation, but as temperature value is increased and the applied stress reduced, the other two processes become increasingly important and will determine the creep behaviour of the metal. Under these conditions the rate of creep is closely related to the rate of self diffusion, that is the rate at which the individual atoms change their position as a result of thermal fluctuations. With increase in temperature, the diffusion rate increases, the material becomes softer and eventually recrystallizes.

The temperature at which this occurs is an approximate guide to the useful practical limit of the material; the temperature is usually about half the melting point on the absolute scale but can be increased by alloying additions allied with appropriate heat treatment.

The creep resistance of a pure metal is usually improved by the addition of elements, that form a solid solution with the parent metal. This is a state, where the atoms of the alloying element are distributed singly in the parent metal either in place of the parent atoms (substitutional) or between the parent atoms (interstitial). The different atomic size of the alloying elements increases the strength by impending the motion of dislocations; it also increases the recrystallization temperature. In general, the bigger the differences in atomic size, the greater the strengthening produced.

Greater improvements in creep strength can be obtained, however, by the addition of finely dispersed particles of the second phase. This is usually achieved by precipitation or age-hardening, making use of an alloying element that shows an increasing solid solubility in the parent metal as the temperature is increased. The alloy is heated to a temperature at which all the alloying element dissolves in the parent metal and it is then cooled quickly enough, in order to retain the alloying element in supersaturated solution. It is then aged by reheating it to a temperature at which the excess solute can precipitate as a second phase. The greater the degree of dispersion the greater the strengthening, but a structure obtained in this way is not stable and the precipitating particles will grow in size if the metal is subsequently subjected to a high enough temperature. To provide a good creep resistance, therefore, the precipitate should be stable to a high temperature, preferably above the recrystallization temperature.

It may occur very often under special operational circumstances, that crack initiation and propagation were found to occur as a result of the formation, growth and linkage of grain-boundary creep cavities and wedge cracks, irrespective of cyclic frequency. Crack growth may occur either by a fatigue or a creep mechanism, or may result from the interaetion of the two processes. The mode of cracking is any instance depends on the temperature, the microstructure of the material and the mechanical properties and parameters that describe the cyclic-loading conditions, such as cyclic fiequency, strain range and hold time. In general, a change from fatigue - to creep dominated behaviour, as manifested by a change from trans - to intergranular cracking, is caused by a decrease in frequency and strain range and by an increase in hold time at a tensile part in the cycle. (2,3)

Accumulation of internal damage

The microstructure of a 2,25 Cr-1 Mo steel embrittled by service in a boiler header by 88000 hours service at $813^{O}K$ was examined with the aid of transmission and scanning electron microscopy. The major microstructural changes due to long term elevated temperature service were evaluated by comparing the steel with a similar grade of virgin steel.

Considerable coarsening and precipitation of carbides had occurred during service, particularly along prior austenite grain boundaries which were covered with thick elongated films of M_6C and $M_{23}C_6$ type carbides, and within the bainitic grains which showed low hardness. M_6C carbide which was not observed in the virgin steel seemed to be predominant in the service exposed one. On the other hand, the dispersions of fine M_2C type carbides seen within the bainite in the virgin steel, were not found in the corresponding areas of the service exposed steel although they were present within the ferrite grains in both steels.

The service exposed steel showed a tendency for intergranular fracture and crack propagation along the interface of the grain boundary carbides and the matrix was observed, suggesting low cohesion at these interfaces. Some grain boundary segregation of phosphorus was confirmed by Auger Electron spectroscopy. In addition, isolated voids of mean size 1,3 μ m were also found on the grain boundaries ofthe service exposed steel. (4)

Thermal strain ageing and crack initiation during low cycle

thermal shock fatigue

Thermal strain ageing is a very complex phenomenon as it affects the metallurgical structure of a material, particularly when thermal effects and plastic deformation are involved. A case in point is the process of thermal shock fatigue. Certain materials can gain strength and creep/fatigue resistance from the presence of dispersed precipitated and/or small particles. The number, size, shape, location and distribution ofthe precipitate may change the stress-rupture, creep and fatigue characteristics. This can influence the life expectancy of the specimens.

Ageing in many high temperature alloys can continue as they may be metallurgically unstable. This is because of the rearrangement of the microstructure toward the equilibrium state at high temperature. Constituents, which are in solid solution, may tend to precipitate causing significant change in the material properties. If the dispersed particles agglomerate and their sizes become larger than an optimum size during service, the hardness of the material and its creep resistance may decrease. This is known as overaging. Small particles may also precipitate in the grain boundaries and this tends to reduce the ductility of the material. These recipitations may occur with or without the application of stresses. (5)

Strain ageing can also occur with some plastic straiñ at moderately high temperatures. Two types of effects can occur: a strengthening of the flow curve and the appearance of one or more jogs in the stress-strain curve. (6)

This leads to a loss in ductility. Dynamic strain ageing occurs in iron and in other body-centred cubic metals and alloys, containing nitrogen, oxygen or hydrogen. This is a consequence of reduction in the average distance between dislocations. This is believed to be caused by the premature immobilisation of dislocations owing to segregation of interstitial solute atoms during deformation. It can take place in the form of carbides, nitrides or other compounds. Once segregation has occurred, the system is in a lower state of strain energy and the dislocations are trapped. (7)

The rate of accumulation of dislocations $d\varsigma/d\varepsilon$ with respect to strain is thus increased proportionately. Here, ς denotes the density of mobile dislocations. A corresponding increase in the work-hardening rate, $dRe/_{d\varepsilon} dE$ prevails where ε is the plastic shear strain and Re the flow stress.

Thermal-stress resistance can also be reduced by chemical attack such as corrosion and/or cyclic oxidation. This can occur when steam, oxygen or other gas come into contact with the material. Oxides, when formed, tend to enhance embrittlement. Discontinuities at the surface layer are developed by cracking of the surface or by the disintegration of a corroded agent that acts as a source of stress concentration. Diffusion may also promote such mechanisms in the interior of the body. Pressure vessel or pipeline materials are examples.

Oxidation and creep damage at high temperature fatigue should be distinguished. (8) Hydrogen, for example, because of its small atomic dimensions of the order of an Angström, can readily diffuse into the grain boundaries and reduce the thermal resistance. This is why so many thermal-fatigue failures are intergranular in nature.

Mechanical work can lead to recrystallization. As grains are broken, energy is stored in the slip planes and in the different grain boundaries. Subsequent heating enhances recrystallization because of the tendency to achieve a state of lower stored energy. Although there is no clearly defined relation between grain size and resistance to thermal-stress fatigue, materials with large grain size generally have lower ductility. This would reduce mechanical and thermal fatigue life and thermal shock fatigue resistance . (9)

In quantifying the failure by fatigue and damage of the material microstructure, the absorbed energy in the specimen plays a role. This has been referred to by L. Gillemot (10) . as the absorbed specific fracture energy (ASFE) which is the energy stored in a unit volume of material. It can be used as a fracture criterion for the nucleation of cracks and related to that of Radon and Czoboly (11) for characterising the size of the plastic zone ahead of the crack. The method has also been extended by Havas et al. (12) For the case of constant strain amplitude loading. F. Gillemot found a relation between the fatigue crack growth data and ASFE (13) which can also be related to the coefficients in the Paris equation, (14) refer to F. Gillemot and E. Czoboly (15) also for the influence of neutron irradiation. Research has been carried out to elaborate on the analytical solution.(16) Experimentally, there appears to be a tendency for the ASFE values of steels to decrease as the number of thermal shock cycles is increased. (17)

Conclusions

Change of microstructure during thermal shock fatigue of the tested Cr-Mo and Cr-Mo-V alloys in the temperature ranges 20-600°C consists of the formation and development of subgrains. After the appearance of a visible network of sub-boundaries, a decrease in subgrain sizes occurs. This is followed by an increase in dislocation density and enhancement of mobility of the structure near the grain boundaries.

It was our experience, for the tested main steam pipeline that high temperature void formation occurred mostly at the edges of small particles after 65000 hours at 540° C and 130 bar when the cyele frequency was below a critical value.(18)
Microcrack nucleation is then followed by the ´growth of micro and macrocracks. Metallurgical changes involved many events such as strain ageing, recovery, recrystallization, grain growth, tempering, precipitation reactions and so on, the sequence of which cannot be clearly determined.

A new failure prevention method

A new and more advanced SQUID sensor system (superconducting quantum interference device) has been developed by Japanese researchers for use in an actual plant. The miniaturised sensor developed for non-destructive measurement is 262 mm high, 152 mm in diameter and 5 kg in weight. Its ability to detect fatigue damage in Type 316 stainless steel has been investigated. The results are summarised as follows.

The sensor can successfully detect fatigue damage in Type 316 austenitic stainless steel. X-ray diffraction experiments showed that the changes in the magnetic field were generated by martensitic transformation in the stainless steel during the fatigue process The increasing trend of the magnetic characteristics of specimens with an increase in the fatigue damage fraction was independent of the strain amplitude in the fatigue test. It became clear, that the new sensor can use these magnetic characteristics to estimate fatigue damage. (19)

Chemical composition of Type 316 stainless steel (wt%)

C	Si	Mn	P	S	Cr	Ni	Mo
0.007	0.48	0.86	0.023	0.002	16.49	11.16	2.13

Failure prevention methods, possibilities of life extension.

Predicting the remaining operating life of plant equipment that has been in service for 15-20 years or more is a serious and actual problem faced by many industries - power plants, chemical companies, metal producers, basic manufacturers.

Engineering materials are often not in thermodynamical equilibrium, therefore they age under high temperature service conditions.

Material damage of pressure vessels and piping system components (mainly elbows and welded joints) is caused by a combination of mechanical loading, high temperature, corrosive environment and their interactions.

In the current economic and political climate, many companies cannot afford to replace existing plants or major components, therefore they must find a proper method to extend their operating life, without decreasing their operational reliability. Accurate, reliable methods for predicting residual life are based on the new results of damage-analysis, new results of material science, revising the construction and on the application of non-destructive online diagnostical methods.

There are possibilities for controlling the microstructure of steam pipe line and pressure vessel alloys, through in proper time applied, intermediate regenerative heat treatments. Determined the low-cycle fatigue and/or creep damage accumulation during a certain service exposure, and provided, that this damage is small enough and is in reversible region, it should be possible to restore. approximately - the original microstructure of the material, and through it the competent material characteristics.

Proper residual life prediction methods and application of revalidation technics permit the continued use of plants that might otherwise be retired from service unnecessarily; and through it one can postpone new investments and it can mean a great amount of money savings for lot of companies.

Possibilities of Life Extension

Some Results of Revalidability Tests

Regenerative heat treatments have been performed with a steam pipe-line material, which operated 163.000 hours at 540 °C, at 105 bar. The pipe-line diameter and wall-thickness values were 325/26 mm. The specified chemical composition is given below in (wt %) Table 2.

Table 2.

Cmax	Si	Mn	Mo	Cr	Pmax	Smax
0.15	0.15	0.4	0.9	2.0	0.04	0.04
	0.50	0.6	1.1	2.5		

We have investigated separately the pipe and elbow as well. Table 3. and Table 4 show some mechanical properties ofthe steel tested in "service exposed" and "after revalidation" conditions.

Table 3.

Absorbed Specific Fracture Energy (ASFE), $W_c (J.c\,m^{-3})$

T ^0C	StressConcentration factor	Straight pipeline		Elbow	
		Service exposed	After revalidati	Service exposed	After revalidation
20	1	923	1446	959	1363
	1,66	793	941	857	1013
	2,15	684	889	846	906
	3,1	363	566	548	648
	4,2	327	530	443	544
540	1	377	948	435	872
	1,66	350	700	433	638
	2,15	326	597	388	549
	3,1	156	207	167	214
	4,2	115	214	155	205

To evaluate the toughness values of the tested material, we measured the Absorbed Specific Fracture Energy (ASFE) as a function of the Stress Concentration Factor (SCF), in the original and the heat treated condition. The measured values are given in Table 3 Table 4. contains the results ofthe standard ISO-Charpy V impact test.

All data are the averages of five measurements.

Table 4.

Charpy-V Impact values, KCV Jcrri [2]

T °C	Straight pipeline		Elbow	
	Service exposed	After revalidation	Service exposed	After revalidation
20	100	238	139	221
0	54	215	96	188
-10	48		55	
-20	35	196	25	
-30	26		18	
-50		94		115
-75		28		25
$t_{tr(KCV=35)}$	$-20^{\circ}C$	$-72^{\circ}C$	$-17^{\circ}C$	$72^{\circ}C$

The damage accumulated during the preceding period was small enough, so it enabled us to reverse the changes that have occurred by means of appropriate, intermediate regenerative heat treatment to restore - approximately - the original mechanical properties of the straight line and the elbow and thereby to increase the effective service life.

It succeeded us to improve significantly the ASFE values, as well as transition temperature values, measured by the ISO-Charpy-V impact test. (20)

References

1.,J.Ginsztler - A.Penninger -L.Szeidl -P.Várlaki "Stochastic Modelling of Stress Processes in Power Plant Boiler Walls" International Journal of Pressure Vessel and Piping (1996), pp. 119-124., Elsevier Science Ltd.

2., P.P. Benham - R.Hoyle: Thermal Stress. p. 280. Sir Isaac Pitman and Sons Ltd, London, 1964.

3., D.Armstrong - G.J.Neate: Crack growth in bainitic O,5Cr-Mo-V steel under creep-fatigue conditions. Materials Science and Technology, January 1985.Vol.1. p.19-24

4., I.Masumoto et al: Electron Microscopic Examination of a Service Embrittled 2,25 Cr-1 Mo Steel. Transaction of the Japan Welding Society, Vol.17. No.2 Oct.1986. p. 36-42.

5., .S.S. Manson, Thermal Stress and Low-Cycle Fatigue, McGraw-Hill, New York, 254 (1966.)

6., J.E. Dorn, Mechanical Behaviour of Materials at Elevated Temperatures, McGraw-Hill, New York, 361 (1961.)

7., A.J. Kennedy, High Temperature Materials. The Controlling Physical Processes, Oliver and Boyd, Edinburgh, 23-24 (1968.)

8., R.P. Skelton and J.I. Bucklow, "Cyclic oxidation and crack growth during high strain fatigue of low alloy steel", Metal Science 12 (2), 64-70 (1978)

9., S.S. Manson, Thermal Stress and Low-Cycle Fatigue, McGraw-Hill, New York, 245 (1966)

10., Gillemot, "Zur rechnerischen Ermittlung der Brucharbeit", Materialprufung 3, 330-336 (1961), L. Gillemot, "Low-cycle fatigue by constant amplitude mean stress", First Internat. Conf. on Fracture, Sendai, Japan, 47-80 (1965), L. Gillemot, " Criterion of crack initiation and spreading", Engrg. Fracture Mech. 8,239-253 (1976))

11., J.C. Radon and E. Czoboly, Proc. Internat. Conf on Mech. Behaviour of Materials, Kyoto, 543 (1972)

12., I. Havas et al. Materialprufung 16, 349 (1974)

13., F. Gillemot, Internat. Conf on Creep and Fatigue in Elevated Temp. Appl. Sheffield (1974)

14., P. Romvari and L. Toth, Gep. 8,281 (1981)

15., F. Gillemot, E. Czoboly and I. Havas, "Fracture mechanics applications of absorbed specific fracture energy: notch and unnotched specimens", Theoret. Appl. Fracture Mech. 4, 39-45 (1985)

16., E. Czoboly, I. Havas and J. Ginsztler, "Relation between low cycle fatigue data and the absorbed specific energy", Fifth European Conf on Fracture, Lisboa, 1-12 (1984)

17., J. Ginsztler and K. Kormi, in: Process Condition Monitoring and Revalidation of Pressure Vessels and Pipe Line Elements Subject to Fluctuating Pressure and Temperature, Leeds Polytechnic Short Course, 1-151 (1984)

18., N.Y. Tang, D.M.R. Taplin and A. Plumtree, "Schema for depicting cavity nucleation during hi temperature fatigue". Mat.Sci. and Technology 1,145-151 (1985)

19., Nonyashi Maeda - Msahiro Otaka - Sadato Shimizu :"Development of an advanced SQUID system for non-destructive evaluations of material degradation in power plants." International Journal of Pressure Vessels and Piping. Vol. 71 No.1. April 1997. p.13-17

20., J.Ginsztler - L. Dévényi: Lecture at the International Conference on "Residual Life of Power Plant Equipment Prediction and Extension." January 23-25,1989, Hyderabad, India.p.4B.5

CRACK INITIATION AND GROWTH DURING THERMAL TRANSIENTS

R.P. Skelton

Imperial College of Science, Technology and Medicine, London, UK

Abstract

This chapter follows the course of a crack in a typical component from the initiation and short crack growth stage, through to the deep crack growth stage and on to the possibility of complete penetration across the wall thickness. The causes of such growth are considered, such as thermal shock and other constraints against expansion or contraction and the many ways of simulating propagation behaviour in the laboratory are discussed, where a cyclic event in service is identified with a fatigue cycle performed in the laboratory. Parameters which are used to describe crack growth in the various regions are explained, together with methods of accounting for internal structural damage in the material ('creep-fatigue interaction') which is observed to enhance crack growth rates. Many worked examples are given, either to illustrate a technical point or based on service experience. Finally, a complete case study (retrospective analysis) of crack propagation across a component in power plant is undertaken together with a validation of the calculations.

1 - INTRODUCTION TO THERMAL FATIGUE

1.1. Background

1.1.1 Origin of Thermally Induced Stresses

Owing to the constraints imposed by thermal expansion and conductivity of metals, heating and cooling transients cause internal stresses between tension and compression in many components, causing yield in the surface layers if the temperature changes are severe. Primary stresses arise from pressures, mechanical end loads ('system stresses') and wholesale constraint, such as a bar prevented from expansion by anchoring both ends. In this case bulk thermal stresses across the section are independent of the rate of temperature change in the elastic region, becoming modified in the elastic-plastic region only when strain rate/relaxation effects arise in the material. Secondary stresses cannot be measured directly. For example, they may arise from the heating and cooling of a dissimilar metal weld (also independent of heating and cooling rates if yield has not occurred). The most common example is one of *thermal shock* [1], where transient stresses arise due to sudden temperature changes at internal or external surfaces of components. In this case the temperature ramp rate is of major importance in determining the magnitude of such stresses.

In assessing the lifetime of components at elevated temperature, it has proved convenient to recognise several *failure modes*, each requiring separate treatment. These may be classified as (i) inelastic deformation (cyclic plasticity), (ii) ratchetting, (iii) creep damage accumulation, (iv) creep-fatigue crack initiation and (v) crack propagation. The latter may also be sub divided into (a) fatigue crack growth under transient conditions and (b) creep crack growth under steady-state conditions. Study of the latter will form a major theme at other contributions in this book. As will be seen, interactions may well occur between the modes of types (i) - (iv). Creep damage is generally regarded as a bulk phenomenon while crack initiation ordinarily occurs at the surface (unless it be from an embedded defect). Cyclic plasticity can either occur in the bulk of a component (i.e. across the whole section) or in a contained zone, depending on component thickness and loading type.

1.1.2 Simulative Tests

Behaviour can be modelled by testing the whole component (sometimes an expensive and unrealistic option) or by replicating one feature of a design ("features" test), and simulating thermomechanical behaviour until signs of distress are evident. Additionally, simple characterisation tests are undertaken on smooth cylindrical specimens to determine *material properties* used in damage assessment e.g. cyclic stress-strain and cyclic relaxation behaviour. Further, for crack initiation calculations, fatigue endurances to some specified failure criterion are also utilised. These low cycle fatigue (LCF) tests often incorporate a tension dwell which simulates the steady running period of power plant or aero-engine. These characterising tests [2, 3] will be discussed later . They must be distinguished from tests to determine *physical parameters* such as Young's modulus, thermal expansion coefficient and thermal conductivity which are a pre-requisite for any stress analysis involving thermal transients. Finally, fundamental experiments are sometimes required which establish a damage parameter, for example the extent of grain boundary cavitation or oxidation during fatigue [4, 5]. Tests involving a mixture of cycles may then be performed in order to verify a particular *model*, for example on creep-fatigue and oxidation-fatigue interaction.

1.1.3 Use of Assessment Codes

As will be shown, it is important to take account of the specimen *size* with respect to the component being modelled. To a large extent, the task of the investigating engineer has been simplified by the use of Design and Assessment Codes. The principal Procedures for use at high temperature are the ASME Code Case N-47 [6] developed in the USA, Code RCC-MR [7] developed in France and the R5 Assessment Code [8] developed in the UK. They have been compared recently [9] for their respective capabilities of predicting crack initiation and growth and will be discussed further below. In essence, these Codes embody over 30 years work of detailed experimentation and the derivation of complex expressions. They can also help avoid a detailed literature search since relevant physical and material parameters for an assessment are often tabulated. Other such data sheets are to be found elsewhere [10, 11].

1.2. Practical Problems in Thermal Fatigue

When massive components such as valve chests, turbine casings and rotors in power plant are subjected to large temperature transients, as during start-up and shut down, thermal fatigue cracking may occur at critical locations. Similarly, thermal shock in thinner section components such as blades and vanes in aero-engines resulting from repeated take off and landing operations may lead to surface crack initiation. It thus becomes important to (i) characterise the initiation and (ii) characterise the growth of such fatigue cracks beneath the surface and to assess their likely depth of penetration with known service cycles and (iii) to determine the possibility of crack arrest.

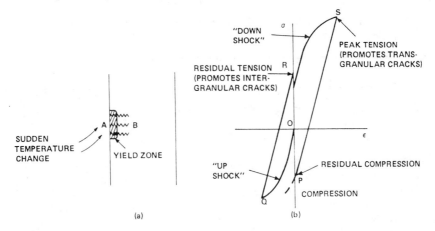

Fig. 1.1 Heating and cooling structure surface (a) leads to hysteresis loop (b)

1.2.1 Thick Component

The generation of thermal stresses may be examined qualitatively as follows [1]. Suppose a thick structure is at a uniformly low temperature and that the temperature suddenly increases at A in Fig. 1.1a. This is an 'upshock' and the surface at A goes into compression as it attempts to expand against the remaining material, yielding along OQ (Fig. 1.1b). Heat now flows slowly towards B, and as the temperature gradient reduces, the whole system expands, taking the surface at A into tension at R, i.e. at the original strain value (approximately). If a period of steady running at high temperature ensues, this residual tensile stress is sometimes able to relax, inducing intergranular cracking in many alloys.

When the temperature suddenly decreases at A, this series of events is reversed. This is a 'downshock' and the surface layers now go into tension as contraction is attempted. Peak tension occurs at point S, promoting transgranular cracking since the corresponding strain rate is high and the temperature has reduced. Later, when the whole structure cools, the yielded section at A is forced into compression by the surrounding metal. The residual compression stress P (again, approximately at the original strain level) does not necessarily create damage, but when many rapid heating and cooling cycles are repeated from point P, the system shakes down to a closed hysteresis loop PQRS at the surface, leading to the initiation and propagation of multiple thermal fatigue cracks due to reversed plasticity.

1.2.2 Thin Section Component

The sequence described above is most often associated with power plant when required to start up and shut down rapidly (temperature range 25-565°C). However, essentially the same cycle occurs in aero-engine components such as blades and vanes, although the time sequence for key events in the cycle is very much reduced. In Fig. 1.2 we have an example of a hysteresis loop [12] experienced by a turbine blade during take off, cruise and landing (temperature range 200-1050°C).

1.3. Examples of Thermal Shock

Table 1.1: Instances of Thermal Fatigue Cracking [1]

Component	Material	Comments
Ingot moulds	Cast iron	Operation above 800°C
Hot rolls	Forged steel,. cast iron	Rolling temperatures 500-900°C. 0.1mm deep cracks
Dies (die casting)	Various steels	Carburization improves life
Forging dies	Steel +3%Mo, Ni	Pre-heating reduces problem
Brake drums	Cast iron	Addition of graphite alleviates problem
Railway wheels	Up to 0.7%C steel	Temperatures up to 540°C attained on braking
Centrifugal casting moulds	21/4CrMo	Craze cracking after 1000 operations up to 600°C
Fast reactor mixing tee	Austenitic stainless steel	Temperature transients up to 250°C in sodium
BWR feedwater system	304 stainless steel	270°C operating, then shut down
Superheater tubes	1Cr1/2Mo, 21/4CrMo, 347 stainless steel	'On load de-slagging' (water quenching)
Inlet headers, ligaments	21/4Cr12Mo, 1Cr1/2Mo	Suspected quench events during start up
Turbine casing	1/2CrMoV	Temperatures up to 550°C
Steam chests	Cast C steel	Temperatures up to 360°C
Valves	Austenitic steel	Temperatures up to 600°C
Turbine rotors	Forged 1CrMoV	Cracks from grooves
Gas turbines (blades, vanes, combustor liners)	Ni or Co base superalloys	Operation between 600°C and 1050°C
Gun barrels	Low alloy steel	Cracking 6mm deep after ~2000 rounds

From the above it may be deduced that (i) components may see either an upshock, or downshock, or both depending on the relative heating and cooling rates, (ii) the magnitude of the surface stress, and hence the amount of yield, depends on the component thickness and (iii) the magnitude of the temperature step change at the surface is not the same as the temperature differential in the metal ($T_s - T_{av}$) that causes the maximum stress during either upshock or down shock. This is examined below.

Examples of thermal shock in service components have been given recently [1] and are reproduced as Table 1.1. An example of craze cracking in a carbon steel economiser recirculating valve is given in Fig. 1.3. The wall thickness was 46 mm and the cracks had penetrated to about 6 mm. It is believed that repeated hot water ingress onto a cool valve was responsible (upshock). However since temperatures were relatively low, cracks were transgranular and there was evidence that they had become dormant.

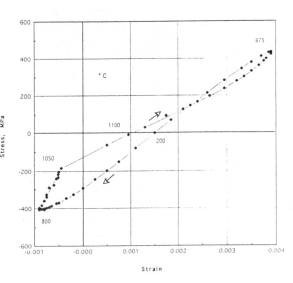

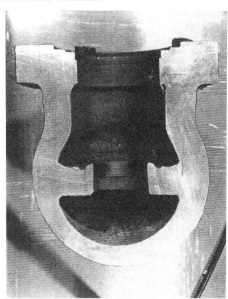

Fig. 1.2 Typical turbine blade hysteresis loop [12]   **Fig. 1.3** Craze cracking in valve bowl [1]

In contrast, a through-wall leak occurred in a 316 austenitic steam pipe at Eddystone supercritical power station in the USA [13]. A cross section of the cracked main steam pipe is shown in Fig. 1.4a. The cause of the premature piping failure was intergranular cracking due to creep rupture, Fig. 1.4b. This failure mode occurred because of residual tensile stresses at the *outside* surface induced in the piping by repeated thermal down shocks at the *inner* surface [13]. The origin of self-equilibrated stresses across the section is considered later.

An example of multiple transgranular cracking at the inside surface of a 1/2CrMoV reheat pipe, which periodically filled up with water, involving some 4350 start up events, is shown in Fig. 1.5 [1].

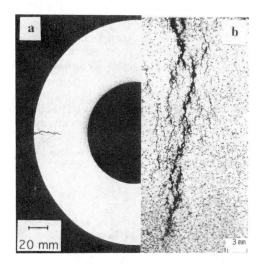

Fig. 1.4 Steam pipe intergranular cracking [13]   **Fig. 1.5** Transgranular cracking [1]

1.4. Laboratory Simulation

Distinction should be made between the *thermal shock* test and the *thermo-mechanical* (TMF) test. The former is usually a simplification of a component shape (wedge, disc, cylinder etc.) relying on internal thermal constraint and temperature gradients to produce a self-equilibrated stress field. In TMF tests, thin hollow specimens are usually employed for excursion rates of 10-20°C s^{-1} in order to avoid temperature gradients [14]. The specimen thus represents a 'volume element' in a component and a variety of temperature/strain combinations can be applied at will ('out of phase', 'in-phase', 'diamond cycling' etc.). The measured stress-strain response is not strictly "mechanical property" data because the yield stress is continuously changing, see Fig. 1.2 for example.

1.5. Useful Relations for Preliminary Lifetime Assessments

Assuming perfect elasticity during a quench or rapid heat event, the peak (equibiaxial) thermal stress generated at the surface is :

$$\sigma = \frac{E\alpha(T_s - T_{av})}{1 - v} \qquad .....(1.1)$$

where v is Poisson's ratio, T_s is the surface temperature and T_{av} is the volume average temperature, E is Young's modulus and α is the coefficient of expansion [15, 16].

To a very good approximation, the maximum value of $(T_s - T_{av})$ occurs after a transit time, t, [17] when the heat wave just reaches the opposite face:

$$t = 0.088 \frac{W^2 \rho c}{k} \qquad \qquad .....(1.2)$$

where ρ is the density of the metal, c is the specific heat and W is the component (wall) thickness and k is the thermal conductivity.

The value of $T_s - T_{av}$ may be calculated from the following expression [18]:

$$\frac{T_o}{T_s - T_{av}} = 1.5 + \frac{3.25}{\beta} - 0.5 \exp\left(-\frac{16}{\beta}\right) \qquad \qquad .....(1.3)$$

where β is the Biot number(= hW/k) where h in turn is the heat transfer coefficient. The term T_o is the initial uniform temperature difference between the body and that of the cooling (or heating) medium i.e. for when a step change in temperature occurs. This expression has been developed by Houtman [19] for calculating (i) temperature differentials in the metal for a linear temperature change at the surface (ii) inelastic strains from elastic strains.

When temperature changes occur inside a body, the inner/outer surface temperature difference, $\Delta T'$, in a cylindrical body at any time t is given by the expression [20]:

$$\Delta T' = \frac{\rho c r_o^2 Q}{2k}\left[\ln\left(\frac{1}{\mu}\right) - \frac{(1-\mu^2)}{2}\right]\left[1 - \exp(-\lambda t)\right] \qquad \qquad .....(1.4)$$

where Q is the temperature ramp rate at the inner surface, r_o is the outer radius, r_i is the inner radius, $\mu = r_i/r_o$, $\lambda = x_i^2 k / \rho s r_i^2$ and x_1 is a function of μ as shown in Fig. 1.6.

During the steady state the exponential term disappears. The ratio of the temperature differential to the inner/outer difference is then given (Fig. 1.6) [20] by:

$$\frac{T_s - T_{av}}{\Delta T'} = \frac{3 - 4\mu^2 + \mu^4 - 4\ln(1/\mu)}{2 - 4\mu^2 + 2\mu^4 - 4\ln(1/\mu)(1-\mu^2)} \qquad \qquad .....(1.5)$$

It can hence be shown [21] that if the cylinder is perfectly lagged, so that under quasi steady state conditions the outer and inner surface temperatures are increasing at exactly the same rate, the axial thermal stress, σ_z, at any radius r is given by:

$$\sigma_z = \frac{E\alpha Q\rho c}{8k(1-v)r_o^2}\left\{3 + \mu^2 - \frac{2r^2}{r_o^2} - \frac{4}{1-\mu^2}\left[\ln\frac{r_o}{r} + \mu^2\ln\frac{r}{\mu r_o}\right]\right\} \qquad \qquad .....(1.6)$$

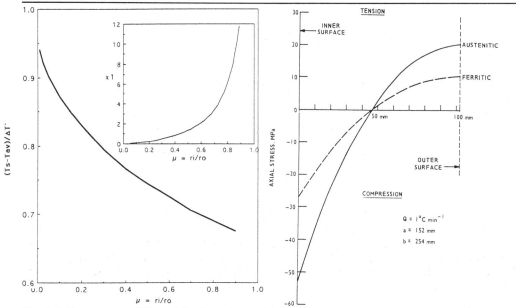

Fig. 1.6 Temperature differential calculation   **Fig. 1.7** Self-equilibrated stress example [22]

1.6. Worked Example [22]

A steam header of outer radius 254 mm and inner radius 152 mm is heated at the inside surface at a rate of 1°C min^{-1}. Compare the thermal stress profiles for the case of a header manufactured out of (i) a ferritic steel and (ii) an austenitic steel respectively. (The following constants may be assumed: $\rho = 7.8$ x 10^3 kg m^{-3}, c = 5.1 x 10^2 J kg^{-1} °C^{-1}, $k = 24$ W m^{-1} °C^{-1} for the austenitic steel and 35 W m^{-1} °C^{-1} for the ferritic steel, E = 1.5 x 10^5 MPa, $v = 0.3$, $\alpha = 1.5$ x 10^{-5}°C^{-1} for the ferritic steel and 2.0 x 10^{-5}°C^{-1} for the austenitic steel.

Substituting relevant values in equation (1.6) and plotting out the stresses across the wall thickness as shown in Fig. 1.7 it is seen that for the given heating rate, thermal stresses generated in the austenitic steel are everywhere about twice those in the ferritic steel. The zero stress corresponds to the position of T_{av} in equation (1.1).

In this problem it is noted that the stress changes sign across the wall (compressive stresses near the bore balanced by tensile stresses near the outside. Also from equation (1.6) it is noted that thermal stresses themselves vary linearly with heating rate.

1.7. Choice of Material

A material's resistance to crack initiation in thermal shock can often be determined by a 'merit order' parameter R, which is defined as:

$$R = \frac{\sigma_y k}{E\alpha} \qquad \qquad .....(1.7)$$

where σ_y is an appropriate (cyclic) yield stress. The higher the value of R, the less is the likelihood of thermal fatigue crack initiation [1]. Typical values of R are given in Table 1.2. Equation (1.7) is thus an approximate guide to thermal fatigue resistance containing both physical and metallurgical parameters. A notable example is the use of single crystals in the [001] direction for gas turbine vanes and blades. Such alloys have relatively high yield stresses and a low value of Young's modulus in this direction [12].

Table 1.2: Typical R values given by equation (8) [1]

Alloy	R, Wm^{-1}
316	1200
Esshete 1250	2450
9Cr1Mo	4600
9Cr1Mo, aged	3600
'Strong 9Cr1Mo' (T91)	4700-5600
12CrMoV	5700-6900
21/4CrMo, normalised & tempered	3900-5200
21/4CrMo, annealed	4000
21/4CrMo, service-exposed	2550
718	5200

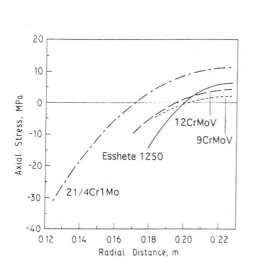

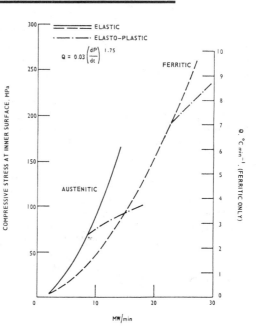

Fig. 1.8 Geometry effect for given transient [21]

Fig. 1.9 Stresses induced on start-up

Surface stresses generated by very severe shocks are dominated by the heat transfer coefficient and are independent of the term k, whilst mild shocks must also take component thickness into account. This has been examined by Rees [21]. In the case of a header of fixed outer diameter 455 mm, four materials were considered in a replacement design, namely one austenitic alloy (Esshete 1250) and three ferritic alloys (9CrMoV, 12CrMoV and 21/4Cr1Mo). Based on long term creep life, the design wall thickness for these materials was 48 mm, 45 mm, 57 mm and 102 mm respectively. By means of equation (1.6) and using 400°C materials data (a reasonable estimate at maximum thermal stress) the stress profiles were calculated and the results are shown in Fig. 1.8 for a Q value of 1°C min^{-1} which kept stresses below yield. An elastic-plastic calculation was in addition carried out for a more severe heating rate of 20°C min^{-1}. The general conclusion in both cases was that, relative to the 9Cr steel, the bore stresses developed in 21/4CrMo cylinders were 7 times greater; for Esshete 1250 the factor was 2.5 while for the 12Cr steel it was 2.

Such calculations can be translated into modified start-up procedures to reduce thermal stresses, arranging for (i) internal gas cooling as is done in many designs of gas turbine blades or (ii) pre-warming of key components [23]. Typical bore stresses induced for a given start up rate expressed in MW min^{-1} are given in Fig. 1.9.

1.8. Use of Assessment Codes

When cracks are discovered during routine overhaul of large components the traditional remedy is either to repair (by excavation and welding), to machine out the defected area as in a rotor [24] or to replace the defected item. The alternative strategy of leaving the component in service until a replacement can be manufactured entails an integrated crack growth calculation which relies on (i) knowledge of temperature gradients generated at the area of interest during heating and cooling cycles (ii) elasto-plastic stress analyses and (iii) relevant high temperature materials data. This is an example of a *remaining* or *extended life* calculation and for safety, pessimistic (i.e. conservative) materials data should be used to predict future behaviour. It may be noted that steps (i) to (iii) may also be applied at the *design* stage, estimating service cycles to crack initiation followed by integrating crack propagation rates from a notional crack depth to some acceptable final depth, assuming stable growth. In this case also, upper bound crack growth rates should be used, based on reliable laboratory data.

In contrast, a *post mortem* or *retrospective* failure analysis would require 'best estimate' crack growth data to explain how a particular failure had come about. This may require further testing of service-exposed material. Care must be taken in applying the Codes [6-8] in this latter mode since safety factors have sometimes been applied to original laboratory data which are not always apparent. The Codes will be referred to from time to time in this chapter. ASME-N47 and RCC-MR contain much physical property and materials data for use in stress analysis while R5 specifically caters for creep-fatigue crack growth in the short and deep crack regimes.

2 - STRESS ANALYSIS AND SUPPORTING LABORATORY DATA

2.1. Introduction

As well as the more familiar assessment Codes such as ASME N-47 [6], RCC-MR [7] and the R5 procedures [8], many organisations have their own in-house design and assessment rules [9]. This section will largely concern the final stages of an assessment where materials properties are closely connected with the engineering input parameters. Before this stage, much preparatory analysis has to be done on bulk structural response, examples being as follows:

- Classify stresses as primary, secondary or bending
- Check for structural shakedown/ratchetting
- Check for significant creep
- Determine elastic follow-up factors
- Determine reference stresses/limit loads

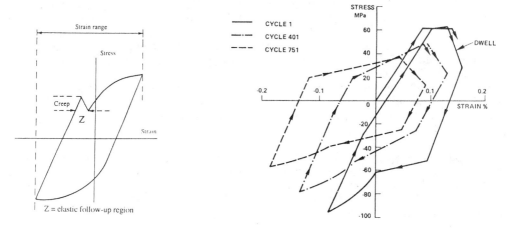

Fig. 2.1 Typical service cycle   **Fig. 2.2** Hysteresis loop evolution in flange [28]

A typical service cycle is shown in Fig. 2.1: its determination might well have involved a full finite element analysis or it might have been deduced by more simple means from available service records. In either case *materials data* are required for (i) determination of the deformation loop itself and (ii) the effect of repetitions of that loop on material/component integrity.

Materials properties may further be subdivided into the well known physical constants (E, α, c etc. as encountered in thermal shock) and other properties depending on *metallurgical* and *environmental* considerations. Examples of these are:

- Yield or specified proof stress values
- Fatigue endurance, with and without dwell
- Creep rupture life and elongation to failure
- Monotonic loading and cyclic stress-strain response
- Stress relaxation properties

Sometimes such data can be found in the Assessment Codes [6-8], usually as smoothed curves or lower/upper bounding lines where unspecified laboratory data have been subjected to a safety factor. Original fatigue data have also been published separately in compendia [10, 25]. However for satisfactory lifetime assessments it is inevitable that additional testing will be called for, sometimes on extracted portions of the service-exposed component.

2.2. Stress Analysis: Contained Deformation

It is important to distinguish between two types of stresses. Thermal transients are examples of *secondary* stresses, i.e. they cause no net force across the component and so are self-equilibrated. Thus an induced tensile stress at the surface due to cooling must be balanced elsewhere in the section by a compressive stress. In contrast , *primary* stresses arise from mechanical sources such as end loads and hoop stresses due to internal pressure. The study of stress-strain fields and their evolution in structures is known as *cyclic inelastic analysis*. The calculations must be pursued until stress redistribution has occurred and stabilisation reached in each volume element of the structure. Complicated finite element calculations requiring constitutive equations are sometimes employed. However simplified methods are available which assume isotropic or kinematic hardening [26]. Alternatively, if the design calculation requires more accurate cycle-by-cycle data, then these may be supplied by direct experiment or computer simulation. Visco-elastic laws and the Bauschinger effect may also describe the process [27]. In either case the gradual change in stress range and plastic strain range are recorded. Figure 2.2 shows the results of a simplified analysis on hysteresis loop evolution in a thin shell/flange feature operating between 494°C and 675°C [28].

In thermal fatigue assessments it will be assumed for simplicity that all components have shaken down to a steady state i.e. there is no further ratchetting. Creep damage is generally assumed to occur in stress relaxation rather than in forward creep. This is the reason why during the last 40 years, high temperature LCF and other related tests have been performed in displacement or strain control.

The link between engineering stress analysis and the tests from metallurgical and materials science is made as follows:

- Calculations are performed and an agreed cyclic *deformation* curve is entered in order to establish a settled total *strain range*
- This strain range is then entered into an agreed *endurance* curve, such as that shown in Fig. 2.3, to arrive at a number of *cycles to failure*.

2.3. Stress-strain Relations

2.3.1 Uniform Stress

A simple but very useful constitutive relation for the hysteresis loop is provided by the Ramberg-Osgood law [29]:

$$\Delta\sigma = A\Delta\varepsilon_p^{\beta} \qquad\qquad .....(2.1)$$

where $\Delta\sigma$ is the stress range (peak tension - peak compression), $\Delta\varepsilon_p$ is the plastic strain range (width of hysteresis loop at zero load) and A and β are constants. Thus a logarithmic plot of $\Delta\sigma$ (from the tips of the loops) versus corresponding values of $\Delta\varepsilon_p$ should yield a straight line of slope β and intercept A at unit strain. Some values of A and β for typical high temperature alloys are given in Table 2.1 [30].

Table 2.1: Stabilised Constants for Cyclic Stress-strain [30]

Alloy	Temperature, °C	A, MPa	β
21/4Cr1Mo, annealed	593	979	0.173
1CrMoV rotor	538	1008	0.090
9Cr1MoV-Nb	538	676	0.110
AISI 1010	538	915	0.228
304/308 weld	482	1330	0.142
AISI 347	600	1354	0.163
"	700	1148	0.185
AISI316	600	1375	0.134
A286	600	4390	0.229
IN617	760	1590	0.140
Inconel 718	650	2896	0.138
Hastelloy X	760	1772	0.215

Table 2.2: Evolutionary Response during Cyclic Stress-strain [30]

Alloy	Temperature, °C	Response
21/4Cr1Mo, annealed	538	H
1CrMoV rotor	538	S
9CrMoV-Nb	538	S
AISI 1010	650	Stable
304/308 weld	593	H
AISI 347	538	H
"	760	H
AISI 316	600	H
A286	600	S
IN617	760	H
Inconel 718	200-650	S
Hastelloy X	538	H
"	760	H

It should be noted that the above parameters are characterised during the *stable state* i.e. after the evolutionary stage where alloys usually cyclically soften or harden. Typical evolutionary data are presented in Table 2.2 [30]. Many analyses require specific early cycle data and this is usually supplied as 1st, 10th, 100th cycle response. There is no agreed method for presenting evolutionary parameters.

The engineering parameter of interest is usually the total strain range, which is the sum of the elastic and plastic strain ranges:

$$\Delta \varepsilon_t = \frac{\Delta \sigma}{E} + \Delta \varepsilon_p \qquad \qquad .....(2.2)$$

and so substituting from equation (2.1):

$$\Delta \varepsilon_t = \frac{\Delta \sigma}{E} + \left(\frac{\Delta \sigma}{A} \right)^{1/\beta} \qquad \qquad .....(2.3)$$

If the total strain range is given, equation (2.3) must be solved numerically for $\Delta \sigma$.

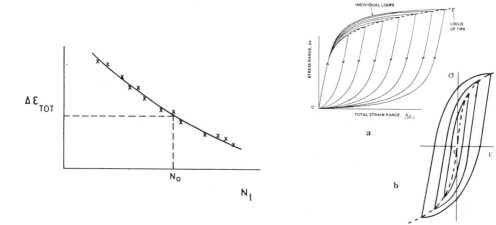

Fig. 2.3 Cycles from endurance curve   **Fig. 2.4** Two ways of plotting hysteresis loops

Equation (2.3) is depicted in Fig. 2.4a as the broken line i.e. with the compressive tips superimposed at one origin. If the loops and corresponding tip loci are nested as shown in Fig. 2.4b, this is equivalent to plotting equation (2.3) in terms of semi range $\Delta \sigma /2$. The value of β is unchanged but the A values must be divided by $2^{1-\beta}$. This plot is useful for comparing cyclic with monotonic loading.

Notes

Other relations are available for describing cyclic stress-strain behaviour [30]. However, advantages of the Ramberg-Osgood relation are:

• The locus depiction is a very good approximation to the actual path taken

- The parameters β and A are coupled such that β/A is generally constant
- Low values of β imply a higher 'yield' stress
- Generally, deformation at low strain rates reduces the A value in a given alloy
- Equation (2.1) is useful for comparing alloys on the basis of a 'cyclic proof stress'. An example is given in Fig. 2.5 [31].
- The von Mises form of equation (2.3) for the equibiaxial stress state is:

$$\Delta\varepsilon_t = (1 - \nu)\frac{\Delta\sigma}{E} + \frac{1}{2}\left(\frac{\Delta\sigma}{A}\right)^{1/\beta} \qquad \qquad .....(2.4)$$

where ν is Poisson's ratio. Equation (2.4) is useful for thermal shock analysis.

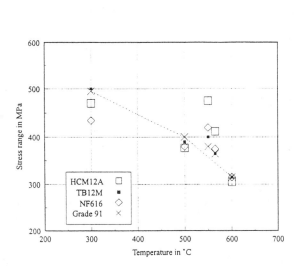

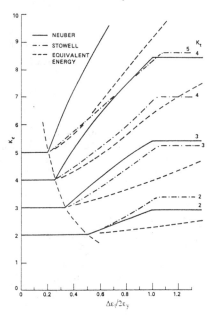

Fig. 2.5 0.05% cyclic proof stresses [31]

Fig. **2.6** Strain concentration factors [30]

2.3.2 Stress and Strain Concentration

In rotors, steam chests etc. many stress concentrators such as grooves, changes in section enhance the strain compared with that occurring remotely [32]. A very simple method of calculating the increased stress or strain at these locations has been given by Neuber [33]:

$$K_\varepsilon K_\sigma = \frac{\Delta\varepsilon_{t\,max}}{\Delta\varepsilon_t} \times \frac{\Delta\sigma_{max}}{\Delta\sigma} = K_t^2 \qquad \qquad .....(2.5)$$

where K_t is the elastic concentration factor, K_ε is the strain concentration factor and K_σ is the stress concentration factor. The subscript 'max' denotes conditions at the root of the groove etc.

Combining equations (2.3) and (2.5) we thus have an expression for the total strain range at the root of the groove:

$$\Delta\varepsilon_{t\,max} = \frac{K_t^2 \Delta\varepsilon_t \Delta\sigma}{E\Delta\varepsilon_{t\,max}} + \left(\frac{K_t^2 \Delta\varepsilon_t \Delta\sigma}{A\Delta\varepsilon_{t\,max}}\right)^{1/\beta} \qquad .....(2.6)$$

Thus, given conditions $\Delta\varepsilon_t$, $\Delta\sigma$ remote from the concentrator, equation (2.6) may be solved numerically to calculate the local (total) strain range. As shown below, the more usual interpretation of the Neuber expression is the intersection of the cyclic stress-strain curve with the hyperbola $\Delta\varepsilon_t\Delta\sigma$ = constant. Other expressions have been derived [30] and in Fig. 2.6 are plotted curves for a rotor steel showing how the strain concentration factor increases with normalised (i.e. to the yield stress) nominal strain range.

It may be noted that equation (2.6) is valid even when $K_t = 1$. This occurs for example for a contained plastic zone in a strain gradient i.e. one surrounded by undeformed material as in thermal shock, even in the absence of a notch [28].

2.3.3 Worked Example

A cylinder of 9Cr1Mo steel at 565°C is thermally quenched at the bore and the peak temperature difference (equation (1.1)) was found to be 286°C. A representative temperature in the yield zone may be taken as 300°C., Assuming $\alpha = 12 \times 10^{-6}$, $E = 1.8 \times 10^5$ MPa, $A = 1881$ MPa, $\beta = 0.193$ and $v = 0.3$ what is the thermally induced strain at the bore?

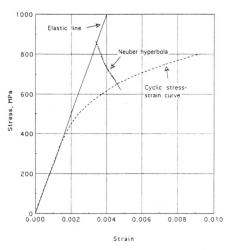

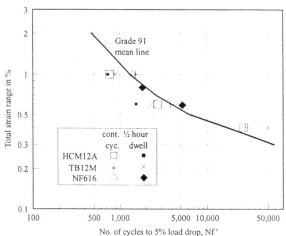

Fig. 2.7 Illustration of Neuber construction   **Fig. 2.8** Initiation from endurance curve [31]

From equation (1.1) the starting (elastic) stress is 865 MPa, corresponding to a strain range of 0.0034. Putting $K_t = 1$ in equation (2.5) we therefore plot the hyperbola $\Delta\varepsilon_t\Delta\sigma = 865$ x

0.0034. This intersects the cyclic stress-strain curve given by equation (2.4) at a strain range of 0.0046 as shown in Fig. 2.7.

The same result may be obtained by combining equations (2.4) and (2.5), putting $K_t = 1$ and noting that $\Delta\sigma = E\varepsilon/(1-v)$ to give:

$$\Delta\varepsilon_{t\,max} = \frac{\Delta\varepsilon_t^2}{\Delta\varepsilon_{t\,max}} + \frac{1}{2}\left\{\frac{E\Delta\varepsilon_t^2}{(1-v)A\Delta\varepsilon_{t\,max}}\right\}^{1/\beta} \qquad .....(2.7)$$

which must be solved numerically to give $\Delta\varepsilon_{tmax} = 0.0046$. This strain range may be entered in an endurance curve , Fig. 2.8 [31], to predict bore crack initiation after about 8000 cycles of repeated heating and quenching.

2.4. Types of Laboratory Data

Data such as those in Table 2.1 are examples of results obtained on a reverse-loading uniaxial testing machine operating in closed loop control, see Fig. 2.9. This type of arrangement, with appropriate heating, rigid grips, X-Y recorders etc. [34] is capable of producing the data listed below in support of lifetime assessments.

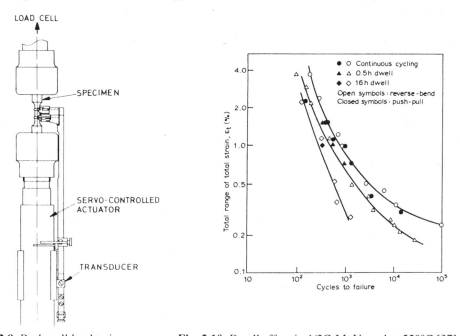

Fig. 2.9 Push-pull load train   Fig. 2.10 Dwell effect in 1/2CrMoV steel at 550°C [37]

2.4.1 Cyclic Stress-strain

Data can arise as the by-product of total endurance tests, values typically being taken at half life. Otherwise, specially conducted block loading tests (incremental step, multiple step) may be undertaken on a single specimen to produce data economically [30]. Energy accumulation methods may be used [35] to guard against crack initiation which would invalidate the data.

2.4.2 Cyclic Stress Relaxation

Relaxation data are required to calculate the creep component of damage experienced while plant is running under steady load. Usually this takes place under constant strain conditions in the laboratory. In service, relaxation can occur under increasing total strain, known as elastic follow-up. The creep rate is defined by:

$$\frac{d\varepsilon}{dt} = -\frac{Z}{E}\frac{d\sigma}{dt} \qquad\qquad .....(2.8)$$

where $Z = 1$ for a constant strain hold and $Z = \infty$ for steady state creep. In service a typical value of Z for cantilever and cylinder-plate applications is about 3.

2.4.3 Fatigue Endurance

The apparatus of Fig. 2.9 is used to determine endurance plots of total strain range against number of cycles to a failure criterion (complete specimen severance or specified load drop, see Section 4). Sometimes the plastic strain range is used since, following the original experiments of Coffin [136], this is considered the damaging parameter, and furthermore gives a straight line on a logarithmic plot:

$$N_f^\gamma \Delta\varepsilon_p = \text{constant} \qquad\qquad .....(2.9)$$

where γ is the slope. For many strong alloys however the plastic strain range becomes very small at quite high values of total strain range.

2.4.4 Fatigue Endurance with Dwell

Depending on the application, creep dwells are inserted in each cycle at (i) peak tension [37], (ii) peak compression [38], (iii) zero mean strain [39] or a combination of these. In low alloy ferritic steels, tension dwells are more damaging whereas the opposite is often reported for superalloys [40]. The data of Fig. 2.10 indicate the typical reduction in endurance that can occur in a 1/2CrMoV steel at 550°C with prolongation of the dwell time [37].

2.4.5 Cyclic Short Crack Growth Data

As discussed in Section 5, short crack growth data can be determined on the set-up of Fig. 2.9, generally using potential drop to monitor the advance rate [3]. Dwells may in addition be introduced. The aim is to reproduce directly in the specimen likely growth rates experienced at the component surface.

2.4.6 Cyclic Deep Crack Growth Data

For this application, totally different specimens such as compact tension or 4-point bending [41] are used, since for cracks propagating more deeply into the component the measuring parameter is the (effective) range of stress intensity, see Section 6. Ingenious arrangements are available for tension-compression cycling with these specimens [41].

2.4.7 Creep Crack Growth Data

Compact tension specimens are again used for the determination of creep crack growth data. In this case however either a steady or interrupted tensile load is applied. This topic is dealt with in other chapters in this book.

2.4.8 Creep Data

Standard creep data are used as part of a damage assessment. The parameter of interest is the variation of ductility with strain rate. Further details are provided in Section 3.

2.5. Analysis of Evolutionary Behaviour

It is finally emphasised that almost all numerical parameters concerning stress-strain response refer to the *stabilised* state. Cyclic inelastic analyses quite often require the stress strain curve to be entered at an earlier state. Exponential-type relations with advancing cycles have been proposed and have been compared [30]. Over a limited initial cycle range, the following relation has been found satisfactory [30, 35]:

$$\Delta\sigma = CN^\lambda \tag{2.10}$$

where the terms C (the initial stress range) and λ are constants. For cyclically hardening materials, λ is positive whereas for those materials which soften, λ is negative. Some examples are given in Table 2.3 [30].

Table 2.3: Evolutionary Parameters for High Temperature Alloys [35]

Alloy	°C	$\Delta\varepsilon_t$, %	C, MPa	λ
316	550	0.6	278	0.114
		1.2	321	0.176
		2.0	378	0.216
9Cr1Mo	550	0.6	630	-0.051
		1.2	675	-0.55
		2.0	683	-0.052
Nimonic 101	850	0.5	776	-0.007
		0.7	989	-0.025
		1.2	1123	-0.039

3 - LINEAR DAMAGE THEORIES OF CREEP-FATIGUE

3.1. Introduction

Our concern is to estimate the number of service cycles to initiate cracking under creep-fatigue conditions. The assumption is that creep and fatigue damage may be evaluated separately and simply summed to obtain the total damage due to all service cycles. The simplest damage techniques use a *linear summation* of the separate effects of fatigue and creep:

$$\Phi_f + \Phi_c = 1 \qquad\qquad .....(3.1)$$

where the subscripts 'f' and 'c' in the total damage factor Φ refer to fatigue and creep respectively. Methods for creep-fatigue lifetime prediction have been reviewed [37, 40]. They include:

- Frequency-modified strain range methods
- Strain range partitioning methods
- Time fraction and ductility exhaustion methods for creep
- Energy expenditure methods [35, 42]

In the following we concentrate on those techniques used in the R5 procedures [8], particularly the ductility exhaustion aspect, but also taking a close look at energy methods. Fatigue damage arises predominantly from cyclic plasticity and is related to the total strain range. Creep damage results from creep strain, either as forward creep or as reversed creep e.g. stress relaxation strain compensated by reversed plastic strain, see Fig. 2.1. Creep damage is uniformly distributed in the bulk whereas fatigue damage is associated with the initiation of a small surface crack. There is no mechanistic substantiation for the linear addition of the two damage types to unity - the evidence is empirical and the assumption is common to most of the models. (There are, in addition, non-linear damage accumulation models and oxidation-fatigue models [5].) Physically, creep strain and the rate at which it is accumulated reflect creep mechanisms giving rise to microstructural damage.

3.2. Scope and Procedure

As originally conceived, the R5 Procedure [8] was intended for ferritic and austenitic steels and their weldments but the initiation procedures could in principle be extended to superalloys operating at temperatures approaching 1000°C. The component or structure is assumed to be defect-free but may contain grooves, notches or other changes in section. The loading may arise from thermal and/or mechanical sources and produce a residual stress state. Each operational cycle is assumed to induce cyclic plasticity and any creep component during steady running may involve a degree of elastic follow-up.

For an assessment, the R5 Procedure follows these logical steps [8, 43]:

- Determine the loading history of the component or feature
- Determine cyclic stress-strain deformation loops
- Obtain creep-fatigue endurances
- Calculate the total creep-fatigue damage, Φ

- Assess whether crack initiation will occur
- Perform a sensitivity analysis
- Report the results

These aspects will be enlarged upon in Section 7. An assessment may involve considerable laboratory testing if data are unavailable in the literature.

3.3. Definitions

As already discussed, the parameters associated with the cyclic stress strain loop are $\Delta\sigma$, $\Delta\varepsilon_t$, $\Delta\varepsilon_p$ as shown in Fig. 3.1a. In addition, if a dwell occurs somewhere in the loop, it is necessary to know the starting stress, σ_0 and the follow-up factor, Z, given by:

$$Z = \frac{\varepsilon'}{\Delta\varepsilon'}$$

.....(3.2)

see Fig. 3.1b. The rate of stress relaxation is given by equation (2.8) viz.:

$$\frac{d\varepsilon}{dt} = -\frac{Z}{E}\frac{d\sigma}{dt}$$

.....(3.3)

see Fig. 3.2.

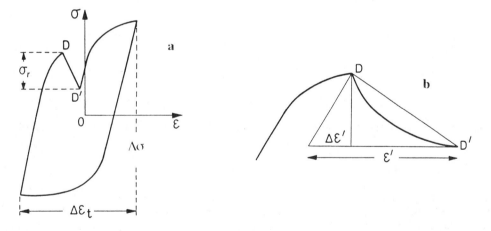

Fig. 3.1 Definition of loop and dwell parameters

A well known equation of stress relaxation is due to Feltham [44]:

$$\sigma_0 - \sigma = b\ln(at + 1)$$

.....(3.4)

where t is elapsed time and a, b are constants. It can be shown that in the presence of elastic follow-up, equation (3.4) is modified to:

$$\sigma_0 - \sigma = \sigma_0 B' \ln\left(\frac{bt}{Z} + 1\right) \qquad \qquad \text{.....(3.5)}$$

The corresponding strain rate is given by:

$$\frac{d\varepsilon}{dt} = \frac{B' \sigma_0 b}{E\left(\dfrac{bt}{Z} + 1\right)} \qquad \qquad \text{.....(3.6)}$$

The relation between initial stress and time, equations (3.4) and (3.5) for a typical material is shown in Fig. 3.3.

Note: Equations other than the Feltham relation may be used.

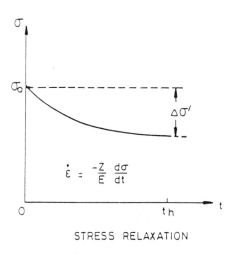

$$\dot{\varepsilon} = \frac{-Z}{E}\frac{d\sigma}{dt}$$

STRESS RELAXATION

$$\frac{\sigma_0 - \sigma}{\sigma_0} = B' \ln\left(\frac{bt}{Z} + 1\right)$$

Fig. 3.2 Establishment of relaxation rate   **Fig. 3.3** Relation between initial stress and time

3.3.1 Fatigue Damage

The fatigue damage per cycle, D_f, is formally defined as the inverse of the continuous cycles to failure at some specified criterion. In the R5 procedures, at each respective strain range, laboratory failure data, N_f, are converted to an earlier failure criterion, N_0, linked to a specific embryo crack size a_0 as defined by the user. Details are given in Section 4. Thus:

$$D_f = \frac{1}{N_0} \qquad \qquad \text{.....(3.7)}$$

3.3.2 Creep Damage

Unless specific long-term creep-fatigue endurances are available [37, 45] which take into account (i) the position and duration of the dwell in the cycle, (ii) elastic follow-up effects (iii) the creep strain rate during the dwell (iv) frequency effects and (v) the effect of prior service exposure, creep damage may be calculated by using existing long term creep failure and ductility data (where the average strain rate is the ductility divided by the time to rupture) and performing stress relaxation tests under cyclic conditions. The traditional damage method uses Robinson [46] time summation of the form $\Sigma t/t_f$, but for creep fatigue problems ductility exhaustion ($\Sigma \varepsilon / \varepsilon_f$) is generally preferred [47].

The creep damage per cycle, D_c, is thus obtained from the expression:

$$D_c = \int_0^{t_h} \frac{(d\varepsilon / dt)}{\varepsilon_f \, \text{fn} \left(\dfrac{d\varepsilon}{dt} \right)} \, dt \qquad\qquad .....(3.8)$$

Equation (3.8) is expressed in an integrated form because quite often ductility decreases as the strain rate decreases, see Fig. 3.4. If compressive dwells are believed to be less damaging, equation (3.8) can be simplified by retaining an upper bound (constant) value of ductility.

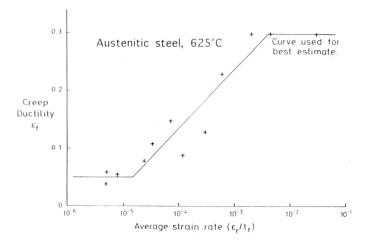

Fig. 3.4 Typical strain-rate ductility plot

3.3.3 Total Damage

The total damage per cycle is the sum of equations (3.7) and (3.8):

$$D_t = \frac{1}{N_o} + D_c \qquad\qquad .....(3.9)$$

so that the creep-fatigue endurance is given by the reciprocal value:

$$N_o^* = \left(\frac{1}{N_o} + D_c\right)^{-1} \qquad .....(3.10)$$

Equation (3.10) is also the basis for damage factors employed in crack growth, see Sections 5 and 6.

3.4. Damage Diagram

From equation (9) the total damage, Φ, resulting from all service cycles of type j is given by:

$$\Phi = \sum_{j} \frac{1}{N_{o,j}} \qquad .....(3.11)$$

If the total damage $\Phi > 1$ then initiation of a crack of depth a_o is predicted.

Alternatively, the creep and fatigue components may be plotted separately on the axes of a Damage Diagram:

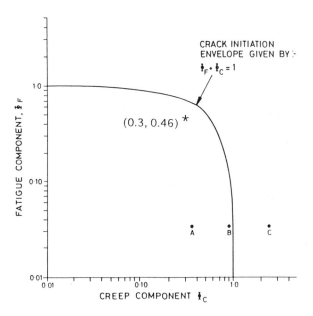

Fig. 3.5 Damage Diagram for assessing creep-fatigue interaction

- The total fatigue damage Φ_f for all service cycles is $1/N_{o,j}$
- The total creep damage Φ_c for all service cycles is $D_{c,j}$

These are plotted as shown in Fig. 3.5. If the point (A, B, C etc.) lies within the envelope, crack initiation is predicted not to occur.

3.5. Worked Example on Creep-fatigue Damage

From a cyclic relaxation test with a 10 h dwell in the laboratory it was established for a low alloy steel at 570°C that the constants in equations (3.5) and (3.6) for a starting stress of 200 MPa were $B' = 0.075$ and $b = 5760$ for strain rate expressed in h^{-1}. Using equation (3.8), estimate the creep damage accrued during the dwell.

If this dwell is combined with a continuous cycle at a strain level where (continuous cycling) crack initiation would occur in 1000 cycles, can the specimen (or simulated component) sustain 300 creep-fatigue cycles without crack initiation?

(Note: The variation in ductility with strain rate (in s^{-1}) may be taken as:

$$\varepsilon_f = 3.18 + 0.38 \log\left(\frac{d\varepsilon}{dt}\right) \qquad .....(3.12)$$

between 10% (0.01) and 83% (0.83) and this is shown plotted in Fig. 3.6. Assume $Z = 3$ for the service location and $E = 1.5 \times 10^5$ MPa.)

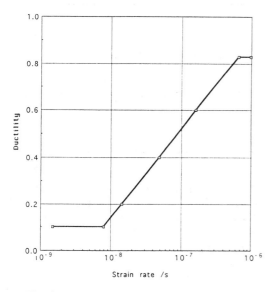

Fig. 3.6 Ductility-strain rate (Worked Example)

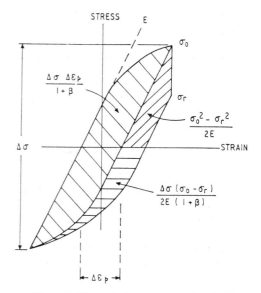

Fig. 3.7 Expended energy partition [35]

For simplicity we divide the dwell period into (i) 10 steps over the first hour (where relaxation is rapid and (ii) 10 steps over the remaining 9 h (where relaxation is more gentle), see Fig. 3.2 for example. We set up Table 3.1 containing the times, the strain rate (equation (3.6)) in that time interval, the corresponding ductility at that strain rate, equation (3.12) and the damage accumulated in the interval given by $(d\varepsilon/dt)/\varepsilon_f \times \delta t$ where δt is 360 s in the first hour and 3600 s thereafter, compare equation (3.8). Incremental damage is also indicated in Table 3.1. By summing these increments, the total damage in the 10 h dwell is found to be 1.52×10^{-3}.

Table 3.1: Creep Damage Calculation

t, h	0.1	0.2	0.3	0.4	0.5	0.6	0.7	0.8	0.9	1.0
$d\varepsilon/dt$, s^{-1} (x 10^{-7})	8.29	4.16	2.77	2.08	1.66	1.38	1.18	1.04	0.93	0.83
ε_f	0.87	0.76	0.68	0.64	0.60	0.57	0.55	0.53	0.51	0.49
Damage increment $\times 10^{-4}$	3.43	1.98	1.45	1.17	0.98	0.87	0.78	0.71	0.66	0.61
t, h	2	3	4	5	6	7	8	9	10	-
$d\varepsilon/dt$, s^{-1} (x 10^{-8})	4.16	2.77	2.08	1.66	1.38	1.19	1.04	0.93	0.83	-
ε_f	0.38	0.31	0.26	0.22	0.19	0.17	0.15	0.13	0.11	-
Damage increment x 10^{-4}	0.40	0.32	0.29	0.27	0.26	0.25	0.25	0.26	0.27	-

For 300 service cycles the fatigue damage is 300 x 1/1000 i.e. 0.3 from equations (3.7) and (3.11). To this we add the total creep damage of 300 x 1.52×10^{-3} i.e. 0.46 and so from equation (3.1) the total damage is 0.76. Creep-fatigue crack initiation is therefore unlikely.

This may optionally be demonstrated by plotting $\Phi_f = 0.3$ and $\Phi_c = 0.46$ in the Damage Diagram of Fig. 3.5 and observing that the point lies within the envelope.

3.6. Accumulated Energy Criterion

One of the earliest ideas in fatigue is that a material will fail when the net work expended (area of hysteresis loop) reaches a critical value. In fact the general damage summation relation of the form $\Sigma N/N_f = 1$ embraced by the R5 procedures is based on the Palmgren-Miner hypothesis [48, 49] which was expressed in energy terms:

$$\sum_{i=1} \frac{w}{W_f} = 1 \qquad\qquad .....(3.13)$$

over each cycle of type i. Miner defined failure as 'the inception of a crack' [49].

Under some circumstances at high temperature, the term W_f has been shown to be a material property, reasonably independent of strain range [35]. Once its value has been established, it may be taken out of the summation sign as follows:

$$\frac{1}{W_f} \sum_{i=1} w = 1 \qquad\qquad .....(3.14)$$

This leads to considerable reductions in experimental testing since W_f does not require to be established at each strain range. There is however some variation with temperature [35]. Typical values of W_f are given in Table 3.2.

Table 3.2: Average Energy Values in Continuous Cycling [35]

Material	Temperature, °C	W_f, Jmm^{-3}
1CrMo0.25V, N&T	482	4.0
1CrMo0.25V rotor	482	2.1
1CrMo0.25V rotor	538	1.8
2.25Cr1Mo, annealed	538	2.1
2.25Cr1Mo, annealed	566	1.2
2.25Cr1Mo, N&T	538	2.0
2.25Cr1Mo, Q&T	482	3.1
9Cr1Mo, N&T	550	3.0
304 stainless steel, ST	566	3.0
304 stainless steel, ST	649	1.6
316 stainless steel, ST	550	0.7
Nimonic 101	850	0.5

It may be shown that the energy expenditure in a single cycle is given by $\Delta\sigma\Delta\varepsilon_p/(1+\beta)$. This may be expressed [4] as:

$$w = \frac{\Delta\sigma^{(1+\beta)/\beta}}{(1+\beta)A^{1/\beta}} \qquad\qquad .....(3.15)$$

If a dwell is imposed at peak tension for example, the extra energy expenditure, δ, is illustrated in Fig. 3.7 and is given by the expression [35]:

$$\delta = \frac{(\sigma_o - \sigma_r)}{2E}\left(\sigma_o + \sigma_r + \frac{\Delta\sigma}{1+\beta}\right) \qquad .....(3.16)$$

Typical values of δ, W_f and w for a rotor steel at 538°C are given in Table 3.3. Thus the extra energy expenditure during the dwell can account for a reduced number of cycles to failure (creep-fatigue interaction).

Table 3.3: Energy Dwell Values for 1CrMoV Steel at 538°C [35]

$\Delta\varepsilon_t$	$\Delta\sigma$, MPa	σ_o, MPa	W_f, Jmm^{-3}	w, Jmm$^-$	δ, Jmm$^-$
0.051	947	363	1.17	4.49 x 10^{-2}	1.11 x 10^{-3}
0.023	825	391	1.22	1.45 x 10^{-2}	0.86 x 10^{-3}
0.012	692	307	1.87	0.42 x 10^{-2}	0.41 x 10^{-3}

It must be noted that the energy values are a measure of heat which has been dissipated in the specimen or volume element of the component. The amount causing damage with each cycle is a very small fraction of this, Further, the model cannot account for the underlying microstructural changes which occur (generally transgranular cracking for continuous cycling and intergranular cracking for creep).

3.7. Validation versus Assessment

Safety against crack initiation can be assured by using:

- Best estimate methodology
- Pessimistic materials data

In retrospective or diagnostic analysis

it is appropriate to use:

- Best estimate methodology
- Best estimate materials data

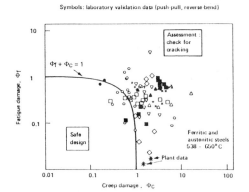

Equation (3.1) is the failure locus on a Damage Diagram, as shown in Fig. 3.8. Thus if laboratory creep-fatigue endurances are plotted and lie wholly *outside* the envelope, then $\Phi_{total} > 1$ and equation (3.1) is a safe bound. Similarly if a retrospective analysis of a failed component also gives a point outside the envelope by applying laboratory data in equation (3.1), then the model has been validated. Conversely, when assessing life at the *design* stage, all future operating cycles/steady running periods should combine to produce a point *within* the envelope for failure avoidance. This is the role of a *sensitivity analysis*. For example, if N_f from laboratory data were replaced by some earlier criterion N_i, the fatigue component Φ_f would move all data points towards unity, so that a point initially within the envelope might end up outside it.

3.8. Closing Remarks

In Section 7 a worked example shows the effect of a sensitivity analysis on a typical result by changing the input parameters if the conditions for crack initiation are violated. Depending on the component, the actions to be taken could include (i) repairing (ii) input data reappraisal (iii) changing operating procedures or (iv) undertaking a crack growth assessment. It is emphasised that two major assumptions of the present lecture have been (i) that isothermal data may be applied to situations where the temperature is changing (ii) that cyclic history effects may be neglected.

4 - PREDICTION OF FATIGUE CRACK INITIATION

4.1. Introduction

Our basic assumption is that significant crack growth cannot be allowed during the lifetime of a component. The purpose of low cycle fatigue (LCF) testing of small specimens is to determine a number of cycles to complete failure, N_f, so that this information can be used for predicting the lifetime. The specimen is taken to represent part of the component and so the term N_f for the small specimen represents crack initiation in a large component. On the other hand, more exacting criteria may be required for critical locations so that knowledge of cycles to crack initiation, N_i, in the specimen itself is desirable. Arbitrary failure criteria may be taken [11] as a 2% or a 5% tensile load drop reckoned from some saturation value, see Fig. 4.1.

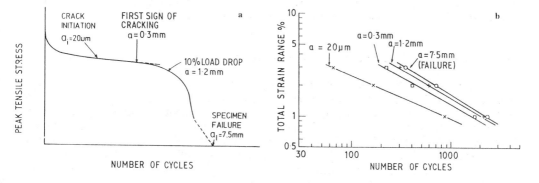

Fig. 4.1 Load drop in fatigue test (a) defines sensitivity of endurance (b)

In Fig. 4.2, ABC is a strain concentrating feature in a component which produces a plastic zone of depth X as a result of temperature or mechanical strain cycling. The effects of a strain gradient in the zone are minimised by making the specimen width or diameter $W \ll X$ (i.e. uniform strain conditions across the specimen). If $X < W$ then either (i) the specimen diameter may be further reduced or (ii) the test may be terminated at a specified load drop, as discussed above. Thus initiation in both "thin" components (e.g. nuclear fuel element cladding) and "thick" components (e.g. the region ahead of a heat relief groove in a rotor) may be modelled [50].

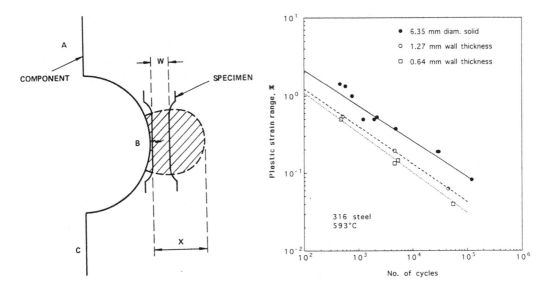

Fig. 4.2 Specimen/component size effect   **Fig. 4.3** Solid/tube specimen size effect [51]

An earlier failure criterion may also be obtained by testing thin components. In Fig. 4.3 is shown the earlier failure of tubular specimens 0.65 mm and 1.27 mm wall thickness compared with failure of standard 6.35 mm diameter specimens [51].

4.2. Fatigue Crack Initiation - the R5 Approach on Size Effect

The R5 approach [8] differs from other Codes [9] in that (i) the value of the damage factor in equation (1) of Section 3 is always set to unity and not < 1 as in some Codes, (ii) no safety factors are applied to fatigue endurance data, assessments being made either on a best-fit curve or a lower bound to the data, (iii) fatigue damage is not applied to the original database (which is generally based on laboratory specimen 'failure' to some specified load drop criterion which contains an element of crack growth) but to an earlier initiation point and (iv) the creep damage summation in equation (3.1) is effected with respect to *strain* (ductility exhaustion), not *time* as noted in Section 3.

The concept has been discussed previously [9]. The relation for estimating initiation cycles, N_o, relies on the propagation characteristics of cracks in the 'short crack growth' regime immediately following initiation (discussed in Section 5). However the crack growth constants do not appear explicitly in the formulae.

Partition of the fatigue cycles is indicated in Fig. 4.4. The term N_1 represents the number of continuous cycles to failure at crack depth a_1 in a laboratory specimen (this could represent 5%, 10% load drop rather than complete severance). It is required to establish the quantity

N_o at a corresponding crack depth a_o. The value of a_o is chosen by the user, having regard to the component. The term N_i in Fig. 4.4 represents the number of cycles to a very early initiation depth a_i (= 20 μm) for use as a fiducial reference, use being made of an empirical relation between N_i and N_l due to Pineau [52] . Along the sloping line in Fig. 4.4 crack growth rate in high strain fatigue is given by [2, 53]:

$$\frac{da}{dN} = Ba^Q \qquad\qquad .....(4.1)$$

where B and Q are empirical constants [2, 53]. It is known that up to a certain crack depth, a_{min}, experimental cyclic growth rates are constant [54], as indicated in Fig. 4.4. In the R5 Procedure a_{min} is taken to be 200μm. In the first Issue of R5 the Pineau relation [52] was modified to :

$$N_i = 0.0366 N_l^{1.306} \qquad\qquad (4.2)$$

but in Issue 2 this has been changed to [55]:

$$N_i = N_l \exp\left(-8.06 N_l^{-0.28}\right) \qquad\qquad .....(4.3)$$

and these relations are shown plotted in Fig. 4.5.

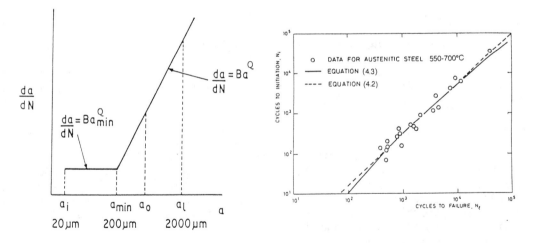

Fig. 4.4 R5 stages in short crack growth   **Fig. 4.5** R5 crack initiation/failure relation [55]

Upon integrating equation (4.1) over the constant- and increasing-crack-growth-rate regions, eliminating the term B and assuming $Q = 1$ (valid for many alloys) then [43]:

$$N_o = N_i + (N_l - N_i) \left[\frac{(a_{min} - a_i) + a_{min} \ln(a_o / a_{min})}{(a_{min} - a_i) + a_{min} \ln(a_l / a_{min})} \right] \quad (4.4)$$

provided $a_o > a_{min}$. A very similar relation applies for $a_o < a_{min}$:

$$N_o = N_i + (N_l - N_i) \left[\frac{a_o - a_i}{(a_{min} - a_i) + a_{min} \ln(a_l / a_{min})} \right] \quad .....(4.5)$$

A very similar result may be derived when $Q > 1$ and is given by:

$$N_o = N_i + (N_l - N_i) \left\{ \frac{a_{min}^Q (a_{min}^{1-Q} - a_o^{1-Q}) + (Q-1)(a_{min} - a_i)}{a_{min}^Q (a_{min}^{1-Q} - a_l^{1-Q}) + (Q-1)(a_{min} - a_i)} \right\} \quad .....(4.6)$$

and when $a_o < a_{min}$:

$$N_o = N_i + (N_l - N_i) \left\{ \frac{(Q-1)(a_o - a_i)}{a_{min}^Q (a_{min}^{1-Q} - a_l^{1-Q}) + (Q-1)(a_{min} - a_i)} \right\} \quad .....(4.7)$$

In summary, the predicted endurance N_o in equations (4.4) and (4.5) (or equations (4.6) and (4.7) respectively) required for a damage calculation depends on the crack size criterion adopted for initiation.

4.3. Other Factors Affecting Continuous Cycling Endurance

It has been shown that a laboratory failure criterion, N_l, must be established from continuous cycling tests so that a value of N_o may be calculated. However some variation in such reference endurances can occur as follows.

4.3.1 Material Scatter

In Fig. 4.6 we plot a typical endurance curve for a service material [45]. There is typically half an order of magnitude variation in endurance at a given strain level. Material scatter may either be *intrinsic* (i.e. invested in the response of the material itself) or *extrinsic* (i.e. arising due to uncertainties of measurement in the testing system, especially strain measurement [56]). Whichever is the case, the assumption of a mean line would give best estimate prediction of behaviour while a lower 95% confidence limit line would be used for a pessimistic or safe prediction.

4.3.2 Frequency Effect

It is generally observed that a decrease in frequency of testing gives rise to a reduction in endurance. An example is given in Fig. 4.7 [57]. For most steels an upper bound may be set at 0.1 Hz at temperatures less than 600°C, but there is in principle no lower bound as the frequency is reduced in continuous cycling. Reduction in endurance is due to creep and oxidation effects.

Note: As shown in Section 3, the effect of creep is assessed separately. There is thus a danger of 'double accounting' if low frequency fatigue data are assumed as a reference but clearly this would be acceptable in a pessimistic prediction.

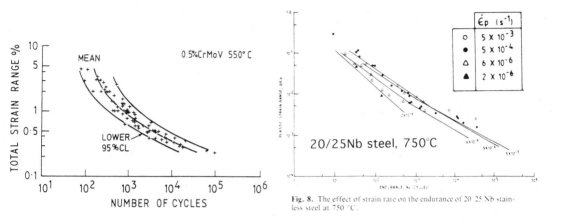

Fig. 4.6 Typical scatter in endurance [45]   **Fig. 4.7** Strain rate effect on endurance [57]

4.3.3 Effect of Temperature

It is generally observed that an increase in temperature gives rise to a reduction in endurance. In ferritic steels for example the reduction factor is about 4 over the temperature range ambient - 550°C. The same factor applies to austenitic steels up to 850°C, see Fig. 4.8 [57]. As a rule of thumb, the effect is small in most steels when the change is limited to about 100°C

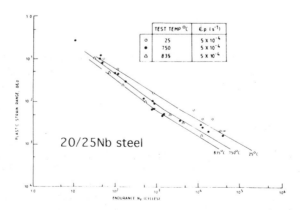

Fig. 4.8 Temperature effect on endurance [57]

4.3.4 Service Exposure

Surprisingly, when plotted on a total strain range basis, the endurance of service-exposed materials lies within the scatter band of virgin material. This has been demonstrated for material taken from a 316 steel turbine valve which had been in operation for 130,000 h at ~600°C subsequently tested at 650°C [58]. A similar effect was found in rotor material exposed to 536°C for 134, 000 h, subsequently tested at 566°C, see Fig. 4.9 [59]. It should be noted however that the cyclic stress-strain properties of service-exposed materials are frequently weaker than that exhibited by virgin material [59].

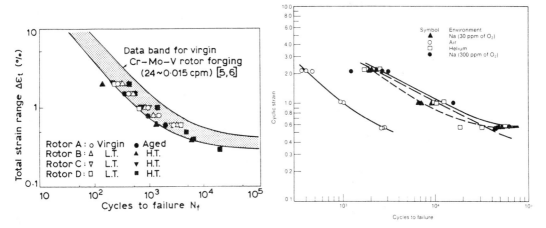

Fig. 4.9 Endurance: effect of service [59]   **Fig. 4.10** Endurance: environment effect [40]

4.3.5 Environment

For fast reactor lifetime assessments, experiments in vacuum, sodium, helium etc. eliminate the oxidation component of fatigue damage. As expected, testing in such environments prolongs the endurance of materials compared with standard tests in air. An example for 21/4Cr1Mo steel at 593°C is given in Fig. 4.10 [40].

4.3.6 Weldments

Weld metal and weldments often show significantly lower endurance than base metal. An example for 316 material is given in Fig. 4.11 [60]. It is believed that this reduction is due to a metallurgical notch effect produced by yield strength variations across a weldment.

4.4. Thermo-mechanical Cycling

As has become clear, the service cycle usually involves a change in temperature and this is sometimes modelled by the thermo-mechanical test discussed in Section 1. The field has been reviewed by Miller and Priest [14]. It is remarkable that for ferritic steels cycled between 300-550°C the data fit well within the broad scatterband of a factor 2 on mean isothermal endurance, see Fig. 4.12. A similar effect is found for 304 austenitic steel in the temperature range 200-750°C as indicated in Fig. 4.13 [14].

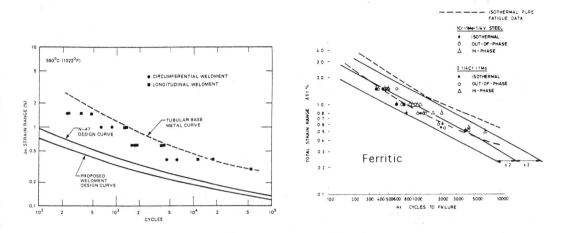

Fig. 4.11 Weld/base metal endurance [60]   **Fig. 4.12** Iso/anisothermal endurance [14]

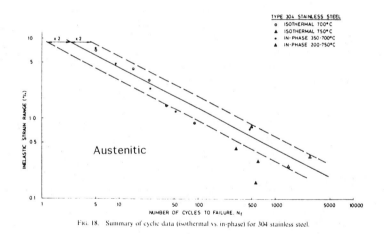

Fig. 4.13 Isothermal/anisothermal endurances

4.5. Validation of Crack Initiation Relations

There are several empirical expressions for crack initiation to specified depths.

$a_0 = 60 \ \mu m$ [61] $N_0 = N_1 - 2.63N_1^{0.9} + 0.69N_1^{1.01}$ .....(4.8)

$a_0 = 100 \ \mu m$ [62] $N_1 = N_0 + 31.52N_0^{0.44}$ .....(4.9)

$a_o = 375 \ \mu m$ [63] $N_o = N_1 - 4N_1^{0.6}$ .....(4.10)

Note: Equation (4.9) must be solved numerically.

Taking $Q = 1$ and using equation (4.2), equations (4.4) and (4.5) as appropriate are validated against the respective equations (4.8)-(4.10) in Table 4.1. Similar comparisons may also be undertaken for (i) Q values > 1 and (ii) using equation (4.3) in place of equation (4.2).

Table 4.1: Comparison of Crack Initiation Criteria for $Q = 1$

N_1	N_o					
	$a_o = 60 \ \mu m < a_{min}$		$a_o = 100 \ \mu m > a_{min}$		$a_o = 375 \ \mu m > a_{min}$	
	Eqn.(4.5)	Eqn. (4.8)	Eqn. (4.4)	Eqn. (4.9)	Eqn. (4.4)	Eqn. (4.10)
200	47	36	57	40	114	104
300	76	73	93	82	176	177
400	110	115	130	131	239	254
500	146	161	170	186	302	333
600	183	209	211	245	367	414
700	221	260	254	308	433	496
800	261	312	298	373	500	579
900	303	366	343	441	567	663
1000	346	421	390	510	635	748
1500	575	715	638	878	984	1178
2000	826	1029	905	1269	1345	1618
3000	1377	1700	1488	2089	2095	2512
4000	1982	2409	2120	2941	2875	3420
5000	2631	3146	2794	3813	3679	4337
10,000	6372	7096	6612	8327	7977	8995

4.6. Worked Example (Sensitivity Analysis)

During a crack growth test, scatter in the data implied that the value of Q in equation (1) lay between the values 1 and 1.45. Assuming the smooth specimen failure criterion (a_1) to be 2 mm after 1000 cycles (N_1), what are the extremes in initiation cycles N_o for chosen values a_o of 60 μm and 375 μm respectively?

To calculate N_i we use equation (4.2) for simplicity. Substituting in the relevant equations (4.4) to (4.7) and taking $a_i = 20 \ \mu m$, $a_{min} = 200 \ \mu m$ as discussed above, we find the following:

To initiate a 60 μm crack 346 cycles are predicted when $Q = 1$ while if $Q = 1.45$, then 363 cycles are predicted. For initiation of a 375 μm deep crack, the predicted cycles are 635 and 735 when $Q = 1$ and 1.45 respectively.

4.7. Worked Example using Energy Criterion

A component is required to operate under mixed cycling according to the scheme in the first three columns of Table. 4.2. The material obeys the Ramberg-Osgood law (equation (2.2)) having the constants $E = 1.5 \times 10^5$ MPa, $A = 1500$ MPa and $\beta = 0.15$. The characteristic energy for crack initiation is 1.0 Jmm^{-3}. Is crack initiation likely to occur in the component?

Table 4.2: Data for Energy Calculation

	Given parameters		Calculated parameters	
Cycle type	No. of cycles	$\Delta \varepsilon_t$, %	$\Delta \sigma$, MPa	W, Jmm^{-3}
Major	50	0.6	593	0.053
Intermediate	250	0.5	554	0.157
Minor	300	0.4	500	0.115
			Total energy consumption =	0.325

Substituting the values of E, A and β in equation (2.2) and solving numerically we have the corresponding stress ranges in column 4 of Table 4.2. The energy in each cycle is given by equation (3.15). Multiplying respective energies by the appropriate number of major, intermediate and minor cycles we then have the accumulated energies for each cycle type in the final column of Table 4.2.

Summing these energies, the total is 0.325 Jmm^{-3} < 1.0 Jmm^{-3} . Crack initiation is thus not likely to occur. This problem is returned to in Section 5.

5 - PREDICTION OF SHORT CRACK GROWTH DURING THERMAL FATIGUE

5.1. Introduction

Crack growth experiments are undertaken to support the contention that structures can perform their design function even though they may contain defects, provided only that cyclic propagation behaviour is well characterised. From the size effect argument in Section 4 (Fig. 4.2) it appears that early growth rates in the component may also be reproduced by monitoring those occurring in the specimen. However it is clear from Fig. 4.2 that crack growth depths must be restricted to depths of approximately $0.25W$, beyond which a notable increase in specimen compliance (inverse stiffness) would be observed. In a thick component, no such increase in compliance would be observed for the same crack depth.

In the following, as in most experimental work supporting lifetime assessments, it is assumed that simulation may be effected by testing at the maximum temperature of exposure, thus giving a conservative (pessimistic) prediction of crack growth rates.

5.2. Form of Crack Growth Law

5.2.1 Empirical Relation

In a typical LCF growth test, a smooth specimen of circular or rectangular cross section is lightly notched and crack progress monitored by one of several techniques listed as under:

*   Potential drop
*   Striation counting
*   'Beach marking'/heat tinting
*   Compliance/load drop changes
*   Surface observation/replica measurements

Testing is normally done between limits of total strain range $\Delta\varepsilon_t$. From such studies an empirical growth law is found [53] to be:

$$\frac{\mathrm{d}a}{\mathrm{d}N} = Ba^Q \qquad\qquad .....(5.1)$$

where $\mathrm{d}a/\mathrm{d}N$ is the cyclic growth rate, B is a constant which depends on the value of $\Delta\varepsilon_t$, a is crack depth and Q is a constant. Values of Q for several materials are summarised in Table 5.1 [2].

Table 5.1: Typical Crack Growth Parameters [2]

Material	Temperature, °C	Q	Notes
20Cr/25Ni/Nb	750	1.0	
316 steel	625	1.0	
High N 316 steel	600	1.0	
Aged 316 steel	625	1.9	1/2 dwell, vacuum
Hastelloy X	760	1.0	
Alloy 800	500-760	1.0	
A286	593	1.0	
9Cr1Mo, annealed	550	1.9	Vacuum
9Cr1Mo, N&T	550	1.3	Vacuum
21/4Cr1Mo, annealed	525	0.74	
21/4Cr1Mo, N&T	525	0.86	
1/2CrMoV	550	1.0	
1/2CrMoV	550	1.5	1/2 h dwell

A typical laboratory test starts with a chordal notch of depth 0.1 mm in a cylindrical specimen of 10 mm diameter and finishes when a significant tensile load drop of 5% or thereabouts has occurred [64]. At this point $a/W \sim 0.25$ depending on crack shape.

For assessing the behaviour of short cracks in a component, therefore, we need to know the total strain range at the surface, $\Delta\varepsilon_t$, see Section 1. The variation of the constant B with total strain in equation (5.1) when growth is expressed in mm/cycle is given in Fig. 5.1 [2]. It may be noted that the value of B is the growth rate at unit crack depth, independent of Q.

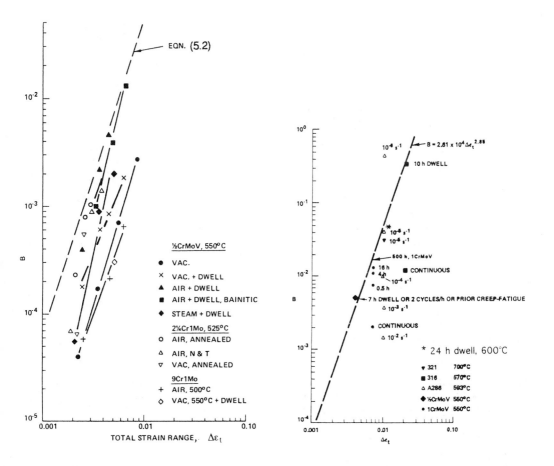

Fig. 5.1 Growth rate variation with strain [2]   **Fig. 5.2** Low strain rate growth data [65]

An upper bound value for B for growth in mm/cycle is given by the relation:

$$B = 2.61 \times 10^4 \Delta\varepsilon_t^{2.85} \qquad \qquad .....(5.2)$$

This relation has been adopted in the R5 Assessment Code [8]. Although intended mainly for continuous cycling data in ferritic steels at temperatures up to 550°C [2], Fig. 5.2 shows that the relation may safely be applied to austenitic steels at higher temperatures and low strain rates [64] and for 316 steel with a 24 h dwell at 600°C. It is also a safe upper bound for very limited data on crack growth in superalloys [9, 53, 65], Fig 5.3. In Fig. 5.4 we have an example [66] of crack growth rates measured during TMF lying well below the upper bound relation of equation (5.2), demonstrating that isothermal data at the maximum temperature are indeed pessimistic.

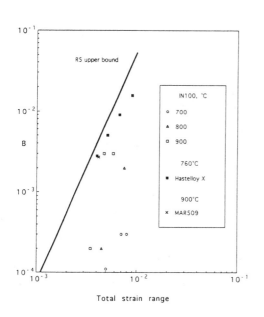

Fig. 5.3 Crack growth in superalloys [9, 53, 65]

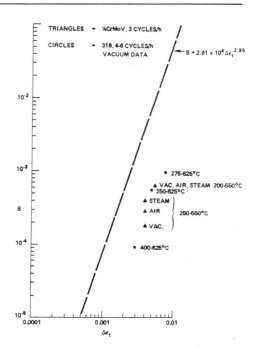

Fig. 5.4 Crack growth during TMF [66]

Note: (i) Equations (5.1) and (5.2) are empirical and, although lying below the upper bound line, crack growth may be either transgranular or intergranular (ii) fundamental theory [67] predicts $Q = 1$ (exponential growth law) in equation (5.1).

5.2.2 Very Low Crack Depths

At very small specimen (and hence component) crack depths ($a/W < 0.01$) equation (5.1) breaks down and growth rates appear to be independent of crack depth for a given strain level. The critical depth, a_{min}, as discussed in Section 4, appears to be in the region of 200 μm. Figure 5.5 shows the effect for an austenitic steel and a superalloy [2]. Abundant data have been provided on austenitic steels [68] and together with detailed studies on short crack behaviour at grain boundaries [69] there may well be a case for reducing the value of a_{min} to 140 μm.

5.2.3 Integration of Growth law

Equation (5.1) may be expressed in its integrated form:

$$N_g = \frac{1}{(Q-1)B}\left(a_o^{1-Q} - a_f^{1-Q}\right) \qquad\qquad .....(5.3)$$

if $Q \pm 1$ or:

$$N_g = \frac{1}{B}\ln\left(\frac{a_f}{a_o}\right) \qquad\qquad .....(5.4)$$

if $Q = 1$, where N_g is the number of cycles involved in propagation, a_o is the initial depth, and a_f is an accepted final depth.

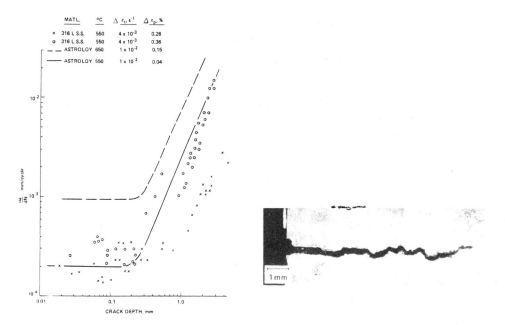

Fig. 5.5 Cut-off at low crack depths [2]   **Fig. 5.6** Intergranular crack in a rotor steel [3]

5.2.4 Worked Example

A component 10 mm thick is required to operate under severe conditions with a surface total strain range of 1%. Assuming a surface defect of 20 μm and that the material obeys an exponential growth law ($Q = 1$) how many cycles are required to propagate the crack to 2.5 mm depth using a conservative growth rate?

Putting $\Delta\varepsilon_t = 0.01$ in equation (5.2), the required value of B is 5.21 x 10^{-2}. The calculation is next performed in two parts. Between 20 μm and 200 μm ($a_{min} = 0.2$ mm) the growth rate is constant. Substituting the value of a_{min} in equation (5.1), this growth rate is 1.04 x 10^{-2} mm/cycle. Over the range 20 μm - 200 μm this represents some 17 cycles. The number of cycles over the range 0.2 mm - 2.5 mm is given by equation (5.4) and is some 49 cycles The total number of cycles is thus 17 + 49 = 66.

Note: If the starting depth (0.02 mm) and final depth (2.5 mm) are instead substituted directly into equation (5.4), the cycle number is overestimated at 93 cycles.

5.3. Industrial Examples on Crack Growth

5.3.1 Crack Growth in a Rotor [5]

Intergranular cracks some 8 mm deep were discovered in a 1CrMoV rotor after some 65,000 h service [70] with steady running periods of 325 h each. Finite element analysis showed that the peak surface strain range was ~0.7%. Simulative LCF crack growth tests with dwells up to 16 h at 550°C showed that the damage mechanism was intergranular [3], Fig. 5.6, similar to that in service [70]. The value of B was given by:

$$B = 8.86 \times 10^{-3} (1 + 0.39 \log t) \qquad \qquad .....(5.5)$$

where t is the dwell time expressed in hours, see Fig. 5.7. Extrapolated values of B out to 300 h lie below equation (5.2). Using equations (5.5) and (5.3), Table 5.2 summarises likely service cycles to propagate a crack up to a final depth of 8 mm assuming an experimentally observed value of $Q = 1.75$.

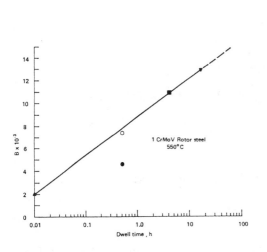

Fig. 5.7 Extrapolation of B to long dwell [5]

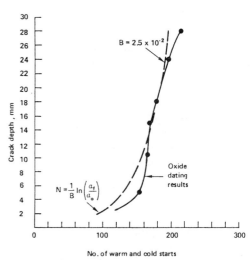

Fig. 5.8 Crack growth in a ligament [5]

Table 5.2: Calculated Cycles to Propagate Crack to 8 mm Depth [5]

Tension dwell, h	Propagation cycles		
	$a_0 = 0.2$ mm	$a_0 = 0.5$ mm	$a_0 = 1.0$ mm
16	321	151	81
100	265	124	67
500	229	108	58

5.3.2 Crack Growth in a Ligament [5]

Interligament cracking in superheater headers has been discussed previously [40]. Transgranular cracking was observed in a ligament area between adjoining 21/4CrMo steam tubes after 85 000 h of service at around 570°C. The unit had experienced some 360 start-ups ('cold', 'warm' and 'hot') during service and the severest thermal loading was estimated to occur in the range 350-450°C. It was possible to perform a retrospective analysis of crack progress using an oxide dating technique [71]. Estimated progress with the number of 'cold' and 'warm' starts (which were considered chiefly responsible) is shown in Fig. 5.8.

The ligament may be regarded as a large laboratory specimen undergoing reversed loading, see Fig. 5.9. A test on service-exposed material in continuous cycling at 400°C and 570°C at a total strain range of 1.3% indicated a transgranular cyclic growth rate, independent of temperature and that $B = 6.6 \times 10^{-3}$, $Q = 1$ in equation (5.1), lying below the upper bound equation (5.2). It may be noted that the best fit of equation (5.4) to the oxide dating line in Fig. 5.8 is by assuming $B = 2.5 \times 10^{-2}$. Separate crack growth tests were undertaken and these indicated that the total strain range in service must sometimes have exceeded 2%.

10 mm

Fig. 5.9 Typical ligament cracking

5.4. Enhancement of Growth Law due to Creep Dwell

5.4.1. Crack Growth Rate

In LCF laboratory tests, potential drop records have shown [3] that crack advance takes place during stress reversal and not in the dwell itself. Thus the increasing values of B with time in Fig. 5.7 are a reflection of the damage accrued during the previous dwell. Metallography confirms that the degree of intergranular cracking is much greater after a 16 h dwell than was the case with a dwell of 1/2 h [3], see Fig. 5.6.

A model has thus been set up [65] using linear damage theory and introducing the term D_c to account for the creep damage per cycle. Under creep-fatigue conditions therefore the basic continuous cycling law given by equation (5.1) becomes enhanced owing to the creep damage accumulated during the dwell. The result when $Q = 1$ is [65]:

$$\left(\frac{da}{dN}\right)^* = Ba\left\{1 + \frac{D_c}{B}\ln\left(\frac{a}{a_o}\right)\right\}^2 \qquad .....(5.6)$$

the asterisk indicating the creep-fatigue modification. A very similar relation applies when $Q > 1$. Values of D_c have been determined empirically for creep-fatigue growth in a variety of alloys [65] and the results may be summarised as a plot of damage factor versus effective frequency [72] where the latter combines the hold time and reversal period, see Fig. 5.10.

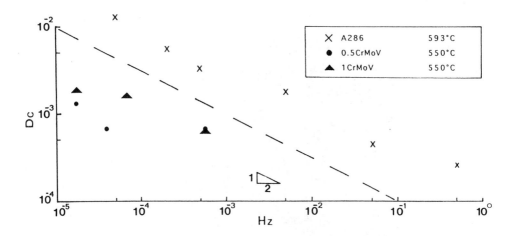

Fig. 5.10 Damage factor extrapolation [72]

5.4.2 Propagation Depth

Equation (5.6) is most useful in its integrated form for block loading over N_i cycles [65].

Continuous Cycling

$$a_f = a_o \exp\left(\sum_{i=1} B_i N_i\right) \qquad .....(5.7)$$

Here the damage function has been set to zero.

Non cumulative Damage

$$a_f = a_o \exp\left\{ \sum_{i=1} \frac{B_i}{D_{ci}} \left(\frac{1}{1 - N_i D_{ci}} - 1 \right) \right\} \qquad .....(5.8)$$

This form appears in the R5 Code [8]. It allows creep damage to accumulate over each block of identical cycles only, resetting the value of D_c for each block. Damage in previous cycles is thus ignored, e.g. when a period of intergranular cracking is followed by transgranular cracking.

True Cumulative Damage

$$a_f = a_o \exp\left\{ \sum_{i=1} \frac{B_i N_i}{(1 - \omega_o)(1 - \omega_t)} \right\} \qquad .....(5.9)$$

where ω_o is all previous cumulative damage up to the current ith block and ω_t is this damage plus that accumulated in the ith block itself.

5.4.3 Worked Example

This case is an extension of the energy worked example in Section 4, now assuming an initial defect size.

A component containing a surface crack of initial depth 0.2 mm receives the loading regime summarised in Table 5.3. Calculate the final depth of penetration for the cases of equations (5.7) - (5.9) respectively.

Table 5.3: Postulated Loading Regime for Component

	Cycles/block	$\Delta\varepsilon_t$, %	D_c	B in eqn.(5.2)
1st application	50	0.6	3×10^{-4}	1.21×10^{-2}
2nd application	250	0.5	2×10^{-4}	7.22×10^{-3}
3rd application	300	0.4	1×10^{-4}	3.82×10^{-3}

This case has been discussed in some detail previously [5, 65]. The results are given in Table 5.4 where the deeper penetration is demonstrated for the two instances of damage. These equations offer a very practical method of engineering assessment and the term D_c can be deduced from metallography [73] or by means of Fig. 5.10 for example.

Table 5.4: Predicted Crack Growth from Separate Damage Models

Type of damage	Penetration depth, mm
None (continuous cycling)	4.88
Non-cumulative	5.41
Cumulative	6.56

5.5. Conclusions

This section has demonstrated that penetration of surface defects may be calculated quite simply from a knowledge of the surface (total) strain range undergone by the component. The advantages of the technique are:

• Generally, no complex finite element analyses or compliance considerations are required for the component

• Appropriate values of the parameter B may be read off from the 'B-$\Delta\varepsilon_t$' diagram, or alternatively an upper bound relation may be used for a conservative assessment

• The relations may only be used to a limited depth ($a/W < 0.25$) in the component, or, if beneath a stress concentrator, to the depth of the calculated plastic zone, see Sections 6 and 7.

6 - PREDICTION OF DEEP CRACK GROWTH IN THERMAL FATIGUE

6.1. Introduction

When small cracks or defects are discovered in structural components, or are postulated to exist at the design stage, an assessment of their likely behaviour during subsequent service cycles must be made. If the crack is surrounded by nominally elastic material, the Paris (LEFM) growth law [74] applies:

$$\frac{\mathrm{d}a}{\mathrm{d}N} = C\Delta K^m$$

.....(6.1)

where a is crack depth, N the number of cycles, C and m are constants and ΔK is the stress intensity range. By integrating equation (6.1) the crack may be progressively updated, i.e. for every known service cycle the incremental growth Δa is added to the starting depth a_0 and the value of ΔK reassessed for the new crack depth, taking into account (i) flaw shape (e.g. extended defect or thumbnail crack) (ii) compliance changes over the propagation range considered and (iii) the nature of the stress distribution. The effect of temperature, dwell (frequency) and environmental effects on the values of C and m must also be taken into consideration.

In materials of very high yield strength it may be shown that the LEFM equation (6.1) may be applied to crack depths down to 0.1 mm at ambient temperature. In contrast, for low yield strength materials the relation breaks down at much larger crack depths, especially at elevated temperatures where materials can become more ductile. It has been shown [2] that cracks less than 3 mm in depth cannot be grown in ferritic and austenitic steels of yield strength 150-200 MPa without substantial reversed plasticity. Similar problems arise with LEFM theory for a crack situated in the plastic zone of a notch.

In principle, the short crack growth equations (5.1) and (5.2) can be used to assess the growth rate of shallow cracks under uniform strain. However the usual service situation involves a *strain gradient* during a thermal transient or in the field ahead of the notch. At some point therefore the plastic strain reduces to zero. This is seen in Fig. 6.1 where curve AA is the theoretical elastic peak profile resulting from a thermal shock at the bore. This is readjusted as the elasto-plastic curve BB. When the temperature becomes uniform the difference between this curve and the original elastic profile define a *residual* stress profile. There is then a period of shakedown for each subsequent shock and the system oscillates between the peak shakedown curve CC and the residual stress state DD. It can be shown that the depth of the plastic yield zone is given by the intersection of the original curve AA with twice the yield stress σ_y as shown in Fig. 6.1. Thus a series of closed hysteresis loops, diminishing in amplitude away from the surface, is described throughout the yield zone during each cycle.

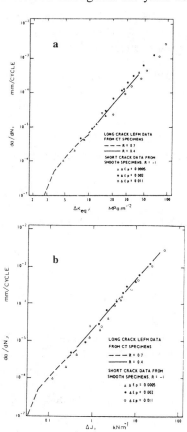

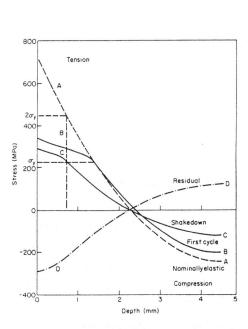

Fig. 6.1 Peak/residual stress system

Fig. 6.2 Alternative data presentation [77]

Several parameters have been developed to correlate crack growth rates in LCF with those in LEFM, as encountered in the plastic zone size transition region of Fig. 6.1. These are either based on a cyclic strain intensity [75] or a cyclic ΔJ [76]. There are marked similarities between the two parameters [77] and a smooth transition can be demonstrated between crack growth rates in the LCF and LEFM regimes [2] as shown below.

6.2. Effective (Equivalent) Stress Intensity Parameter

The equivalent stress intensity ΔK_{eq} is given [75] by:

$$\Delta K_{eq} = (E\Delta\varepsilon_p + q_0\Delta\sigma)Y\sqrt{\pi a} \qquad .....(6.2)$$

where $\Delta\sigma$ is the stress range, and Y is a compliance function which may be taken as unity for a rigidly-gripped single-edge notch specimen or $2/\pi$ for a thumbnail crack. It has been demonstrated [77] that the method is in good agreement with the more familiar formula of Shih and Hutchinson [78] for estimating ΔJ values for short cracks.

Note: beyond $a/W = 0.25$ the value of Y increases above unity as FE calculations have shown [79]. Experimentally , Y is also found to depend on the constraints of the specimen gripping arrangement.

The Paris law, equation (6.1), now becomes:

$$\frac{da}{dN} = C\Delta K_{eq}{}^m \qquad .....(6.3)$$

Some workers prefer to plot their results in terms of ΔJ where:

$$\Delta K_{eq} = \sqrt{\Delta J E} \qquad .....(6.4)$$

A general growth law is thus alternatively expressed as:

$$\frac{da}{dN} = C'\Delta J^{m'} \qquad .....(6.5)$$

In Fig 6.2a, b are plotted some data for EN16 steel plotted according to equations (6.3) and (6.5).

For deep cracks, the expression of Dowling [80] for centre cracked plates may be modified (by the omission of a factor 2) for edge-notched specimens to give:

$$\Delta K_{eq} = \left\{ q_0^2 \Delta K^2 + \frac{q_0 \Delta P \delta_p E}{(1+\beta)(W-a)} \right\}^{1/2} \qquad .....(6.6)$$

where ΔK is the elastically calculated value, δ_p is the plastic displacement, discussed below, and β is the cyclic hardening coefficient discussed in Section 2. The agreement of equation (2) for crack growth in the region 0.2 - 5 mm and equation (6.6) for growth in the region 5 - 17 mm for typical specimens and values of β is shown in Fig. 6.3 [2].

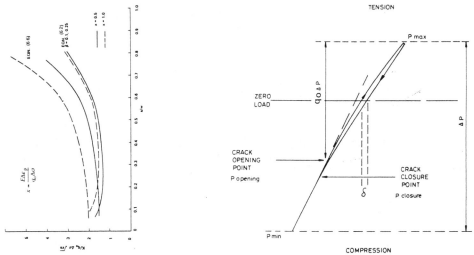

Fig. 6.3 Comparison of growth formulæ [2]   **Fig. 6.4** Typical crack closure [81]

6.3. Crack Closure Effect in Reversed Loading

Evidence for crack closure in laboratory specimens is provided by load-displacement loops and an example is given [181] in Fig. 6.4. There is a clear change in stiffness (compliance) as the crack opens and closes in a single cycle. The important point is that this occurs whilst the specimen is still in the compressive phase. Since the experiment of Fig. 6.4 was performed in displacement control simulating the service condition it is likely that the effect will also occur in service.

It is thus required to define a load range describing the stress intensity range ΔK. Following a recent round robin series of tests [41] it was shown that the best estimate of the *opening range* will provide the most physically acceptable value of ΔK.

The opening parameter q_0 is thus defined by:

$$q_0 = \frac{P_{\max} - P_{\text{opening}}}{P_{\max} - P_{\min}} \qquad \text{.....(6.7)}$$

where P is the applied load, see Fig. 6.4. The effective stress intensity parameter ΔK_{eff} is thus given by:

$$\Delta K_{\text{eff}} = q_0 \Delta K \qquad \text{.....(6.8)}$$

where $\Delta K = K_{\max} - K_{\min}$.

It has been shown by examination of a large body of data [82] that the parameter q_0 varies with crack depth and hence with the term R where:

$$R = \frac{P_{\min}}{P_{\max}} = \frac{K_{\min}}{K_{\max}} \qquad\qquad .....(6.9)$$

Further, it has been shown that an upper bound to the data is given by:

$$q_0 = \frac{1 - 0.5R}{1 - R} \qquad\qquad .....(6.10)$$

A typical example is given in Fig. 6.5 [82]. Equation (6.10) is used in the R5 procedure [8]. Diagrams such as Fig. 6.5 for different materials thus enable an estimate of crack opening ratio to be made from stress (intensity) ratios occurring in service.

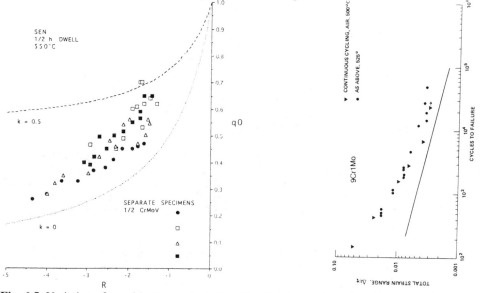

Fig. 6.5 Variation of q_0 with stress ratio [82]   **Fig. 6.6** Lower bound endurance prediction [84]

6.4. Integration of Growth Laws

Equations (6.2) and (6.3) may be integrated to give:

$$N_g = \frac{2\left(a_0^{1-m/2} - a_f^{1-m/2}\right)}{C(m-2)\left\{Y\left(E\Delta\varepsilon_p + q_0\Delta\sigma\right)\sqrt{\pi}\right\}^m} \qquad\qquad .....(6.11)$$

This relation gives the number of cycles N_g to progress the crack from depth a_0 to final depth a_f. Making use of equations (2.1) and (2.2), equation (6.11) may be expressed entirely in terms of total or plastic strain range. As such, this gives a lower bound on smooth specimen endurance since crack initiation cycles are neglected [83]. An example [84] is given in Fig. 6.6.

6.5. Effect of Oxidation and Dwell

It is known that both oxidation and creep effects accelerate cyclic crack growth rates with respect to those observed in continuous cycling [5]. The relation is usually written in separate components:

$$\left(\frac{da}{dN}\right)_{tot} = \left(\frac{da}{dN}\right)_{mech} + \left(\frac{da}{dN}\right)_{creep} + \left(\frac{da}{dN}\right)_{oxid} \qquad \text{.....(6.12)}$$

where the reference 'mechanical' rate may be determined under vacuum [85]. Some workers combine the last two terms of equation (6.12) into a single contribution [5]. For practical assessment purposes, the enhancement of growth due to creep-oxidation-fatigue interaction is reflected by the empirically determined value of C and m in equations (6.1) and (6.3) for example.

Note: equation (6.12) applies equally to LCF crack growth but again this is accounted for by empirical determination of the constants B and Q in equation (5.1).

Similarly, the damage factor, D_c may also be employed in the deep crack regime. In this case however, damage is restricted to the *cyclic plastic zone* at the crack tip rather than in the whole section of interest [65, 72]. This plastic zone size, r_p, given by :

$$r_p = \frac{\Delta K^2}{4\pi\sigma_y^2} \qquad \text{.....(6.13)}$$

where σ_y is a representative yield stress, must be distinguished from the plastic zone due to general yield conditions at the surface, see Fig. 6.7 for the case of a thermally shocked cylinder. The corresponding form of equation (5.6) is:

$$\left(\frac{da}{dN}\right)^* = \frac{C\Delta K^m}{2}\left\{1 + 2M + (1+4M)^{1/2}\right\} \qquad \text{.....} \quad (6.14)$$

$$\text{where} \quad M = \frac{D_c}{4\pi C\Delta K^{m-2}\sigma_y^2}$$

Again, the basic continuous cycling law given by equations (6.1) or (6.3) becomes enhanced owing to the creep damage accumulated by the dwell. Typical values of D_c for a range of high temperature alloys are given in Fig. 6.8 [72].

The model expressly assumes that creep crack growth does not occur during the dwell itself but occurs during the ramp, enhanced because of damage set up in the previous dwell. Experimental evidence for this is provided in Fig. 6.9 [3]. Regarding sequential cycles of type j etc., the value of D_c is continuously updated as was done for the case of short crack growth (Section 5). In this case however, the possibility of overlapping cyclic plastic zones must be taken into account [65].

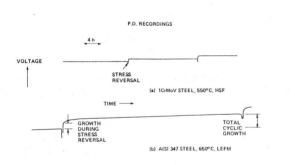

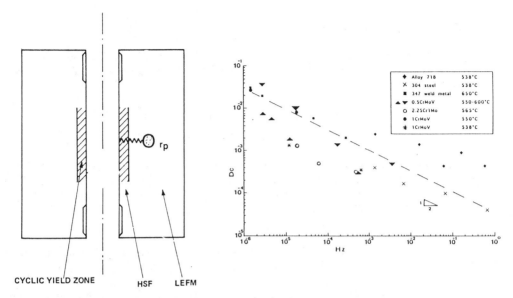

Fig. 6.7 Surface and plastic yield zones

Fig. 6.8 Damage factor extrapolation [72]

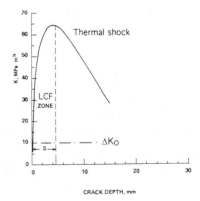

Fig. 6.9 P.d. indicator of crack growth [3]

Fig. 6.10 Peak stress intensity profile

6.6. Multiple Cracking and Crack Arrest

In order to carry out a crack growth assessment on a component or feature, a representative stress intensity range profile is required under peak loading conditions. This can be achieved by weight function methods [86] or finite element analysis. For simple geometrical shapes, analytical solutions are available in the presence of arbitrary loading. Thus for a thick cylinder quenched at the bore, the peak stress intensity is given by [1]:

$$K = \frac{E\alpha(T_s - T_{av})}{1 - v}\left\{1 - \frac{16a}{3\pi W} + \frac{2}{3}\left(\frac{a}{W}\right)^2\right\} \qquad \dots\dots(6.15)$$

where the parameters have been defined in Section 1.

It is emphasised that equation (6.15) does not apply for relatively thin-walled geometries for which numerical analysis is required. For a given wall thickness and thermal transient, the peak stress intensity increases with increase in shell diameter.

In Fig. 6.10 is shown the peak stress intensity profile for a thick cylinder (92 mm outside diameter, 20 mm inside diameter) water quenched at the bore from 550°C [15]. When there is multiple cracking, the net effect is to reduce the stress intensity [1]. The general characteristic of stress intensity profiles for thermal shocks is an initial rise, passing through a maximum, then a decline. This introduces the possibility of crack arrest at a depth defined by the threshold stress intensity ΔK_0, see Fig. 6.10 and this has been demonstrated experimentally [15]. When an end load is present however, the arrest point is pushed further and further towards the back face so that at some point crack breakthrough can occur [1, 87].

6.7. Industrial Example of Crack Growth in a Steam Chest [5]

Post-mortem metallographic examination of a weld feature in a low alloy ferritic steel steam chest of 100 mm wall thickness showed that a crack had propagated along the fusion boundary about 44 mm from the bore over a period of 430 start-up cycles. The peak transient stress profile and the on-load residual stress profiles were converted into corresponding profiles of stress intensity factor, using a computer program based on the weight function method [86]. As shown in Fig. 6.11 the peak transient profiles are compressive whilst the residual profiles are tensile.

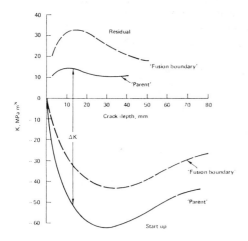

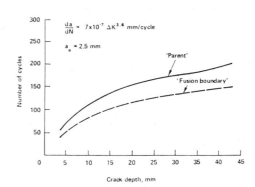

Fig. **6.11** Profiles in thick component [5]   Fig. **6.12** Integrated crack growth rates [5]

Determining the R and q_0 value at each stage by means of equations (6.9) and (6.10) and hence the corresponding stress intensity range from equation (6.8), crack growth rates were determined by means of equation (6.3) with $C = 7 \times 10^{-7}$ and $m = 3.6$ as an upper bound (in mm/cycle) [81]. Integrated crack growth rates are shown in Fig. 6.12. They provide a pessimistic prediction of crack progress and also a sensitivity to the cyclic stress-strain properties of parent material and weld boundary material.

6.8. Worked Example in LEFM Growth with Damage [65]

Experiments on a high temperature steel established that continuous cycling crack growth rates were given by equation (6.1) with $C = 6.18 \times 10^{-7}$ and $m = 2.20$ for growth in mm/cycle and a typical yield stress at high temperature was 150 MPa. Using the expected service application below, determine crack progress.

- (1) 50 MPam$^{1/2}$ for 50 cycles at a damage level $D_{ca} = 6 \times 10^{-3}$
- (2) 40 MPam$^{1/2}$ for 500 cycles at a damage level $D_{cb} = 3 \times 10^{-4}$
- (3) 60 MPam$^{1/2}$ for 40 cycles at a damage level $D_{ca} = 6 \times 10^{-3}$
- (4) 30 MPam$^{1/2}$ for 100 cycles at a damage level $D_{cb} = 6 \times 10^{-3}$
- (5) 40 MPam$^{1/2}$ for 100 cycles at a damage level $D_{cb} = 6 \times 10^{-3}$

The integrated version of equation (6.14) is simply:

$$\delta a_j = \frac{C \Delta K^m}{2}\left\{1 + 2M + (1+4M)^{1/2}\right\}N_j \qquad \qquad .....(6.16)$$

In principle, we substitute the relevant damage levels into equation (6.16) at each step, work out the incremental crack growth, and sum these to give the total crack penetration. However there is a complication due to 'overlapping zones'. The working is shown in Table 6.1 and illustrated in Fig. 6.13. Step 1 is straightforward. The second step is notionally at the reduced damage level D_{cb}, but the zone depth must initially be traversed at the previously inserted level of D_{ca}. After 8.84 mm of growth, the cyclic damage level abruptly changes to D_{cb} for the remainder of this step.

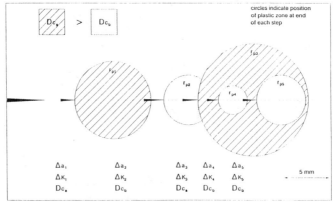

Figure 14   Example of crack growth sequences during LEFM cycling

Fig. 6.13 Overlapping zones in LEFM [65]

Table 6.1: Summary of Damage Levels and Associated Crack Growth

Step	ΔK, MPam$^{1/2}$	Damage /cycle	δa_j, mm	r_p, mm	N
1	50	6×10^{-3}	3.41	8.84	50
2	40	6×10^{-3}	8.84	8.84	204
			1.44		+ = 500
		3×10^{-4}		5.66	296
	60	6×10^{-3}	3.95	12.73	40
4	30	6×10^{-3}	2.42	3.18*	100
5	40	6×10^{-3}	4.34	5.66*	100

*   Not effective, due to overlap of previous zone

The third step reverts to damage level D_{ca} and produces the largest plastic zone size of all, which encompasses the two smaller zone sizes calculated for steps (4) and (5). The high damage level D_{ca} is thus maintained for the rest of the calculation.

The total growth increment over the whole interval is thus $\Sigma \delta a_j$ i.e. 24.4 mm. This would be a conservative calculation since in practice overload effects can cause a delay before crack propagation resumes.

7 - VALIDATION AND WORKED EXAMPLE - CRACK INITIATION & GROWTH IN THICK COMPONENT

7.1. Introduction

In this detailed example we draw together many of the aspects covered in Sections 1 to 6. It concerns the failure of an actual steam chest which had been in service at 570°C peak temperature in power plant and the method closely follows the procedures on creep-fatigue initiation recommended in Volume 3 of the R5 Assessment Procedure [8]. The component appeared to have failed across the whole section by intergranular creep-fatigue crack growth accompanied by grain boundary cavitation.

7.1.1 Component

This was a control valve in a steam line. Its shape can be approximated by a cylinder of outside diameter 800 mm having a wall thickness of 155 mm and internal height 300 mm. Cracks occurred at a change of section at the corners of the vessel.

7.1.2 Heat Treatment

Before entering service, the 1CrMoV component had been normalised at 1036°C followed by an oil quench. It was subsequently tempered for 10 h at 704°C giving a bainitic microstructure of grain size 200 - 400 μm.

7.1.3 Operating History

The operating history at the time of failure was as follows:

- Hours run   70,000
- No. of hot starts   106
- No. of cold starts   127
- ∴ total no. of starts   =   233
- ∴ average "dwell" time   =   300 h

It is required to predict the number of service cycles to initiate a crack 9.5 mm deep (i.e. within the surface plastic zone size) in a wall which is 155 mm thick.

7.2. Scope

7.2.1 Requirements

There are 7 steps in the procedure. They require service data, elastic stress analyses for peak transients, materials properties, elastic-plastic stress analyses and damage calculations.

Note: Many of the calculations were performed on *material-specific* properties. These were obtained on specimens machined from material from the failed chest and included creep, fatigue (continuous cycling) endurance, cyclic stress-strain and cyclic stress relaxation tests.

The 7 steps of the procedure are as follows:

- (1) Obtain the service load history and perform stress analysis of significant cycle types
- (2) Obtain cyclic stress-strain loops including creep during dwell
- (3) Obtain cyclic endurances relevant to initiation depth, a_0, cyclic stress-strain data, relaxation properties etc.
- (4) Calculate service cycles to initiate crack (to depth of surface plastic zone only) and express result in terms of total damage, Φ
- (5) Plot result on Damage Diagram (optional)
- (6) Perform sensitivity analysis with respect to (i) service cycle type and (ii)   materials properties
- (7) Report results of the assessment

7.2.2 Key Definitions

- a_0 Initiation crack depth (variable, incorporates size effect). Changed to a_c (suffix 'component') in this analysis since a_c > specimen width

- N_o Continuous cycling endurance corresponding to a_o. Changed to N_c to correspond with a_c defined above
- D_c Creep damage per cycle (using ductility exhaustion). A bulk failure mechanism
- N_o* Required parameter (creep-fatigue endurance)

7.3. The Seven Steps

7.3.1 Step 1

In Fig. 7.1 is shown, from service records, the inner surface metal temperature rise during a typical start up together with the time of the peak transient. In this case the peak through-wall temperature differential was calculated to be 200°C. In other cases, the severity of the starts according to time off load was defined as 'major', 'intermediate' and 'minor' cycles as shown in Fig. 7.2. Typical peak elastic transients calculated from finite element analyses are shown in Fig. 7.3.

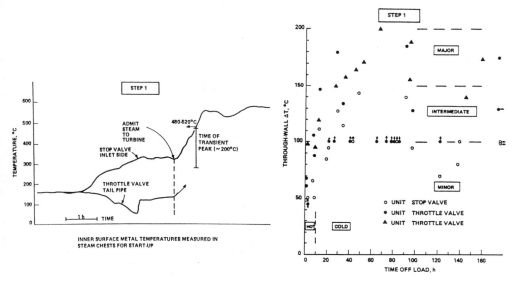

Fig. 7.1 Data from service records   **Fig. 7.2** Start-up temperature differentials

A smaller profile of opposite sign also occurred during cooling transients when coming off load. This is shown in Fig. 7.4. To this profile is added the hoop stress due to steam pressure, giving a resultant stress profile, Fig. 7.4.

7.3.2 Step 2

In order to obtain typical hysteresis loops experienced at the component surface during service, cyclic stress relaxation and cyclic stress-strain data are required. A typical relaxation result obtained from steady-state cycling of the bainitic alloy at 565°C is shown in Fig. 7.5. Similarly, typical cyclic stress-strain results for two service strain rates at 500°C are given in Fig. 7.6.

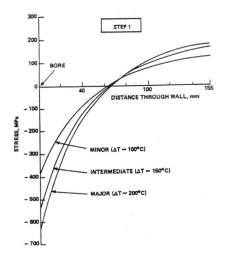

Fig. 7.3 Start-up peak thermal stresses

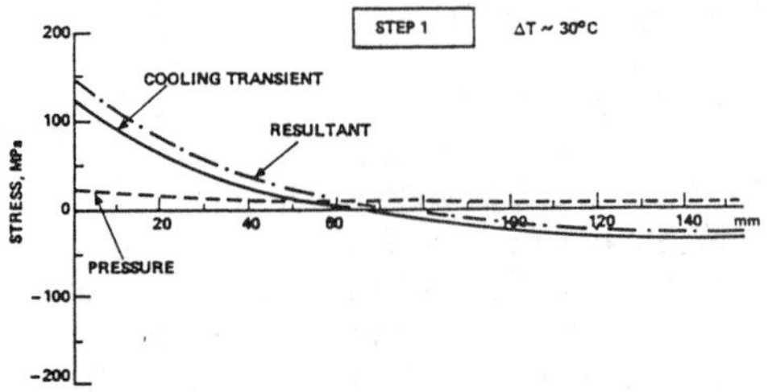

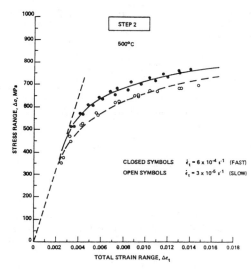

Fig. 7.4 Peak stresses on cooling

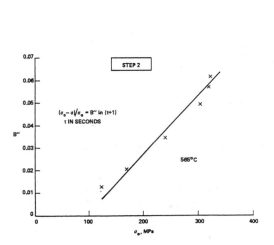

Fig. 7.5 Cyclic stress relaxation data

Fig. 7.6 Cyclic stress-strain data

Using the Neuber relation [33] for a stress concentrating feature as discussed in Section 2, the compression-going and tension-going attributes of the loop were calculated using the data in Figs 7.5 and 7.6 for example. These calculations were pursued cycle by cycle until loop closure (shakedown) had been demonstrated. An example for a major cycle is shown in Fig. 7.7. Summarising data for other cycle types are given in Table 7.1.

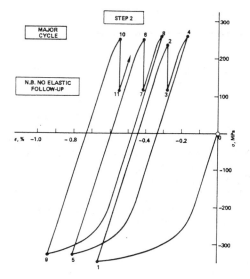

Fig. 7.7 Hysteresis loop shakedown

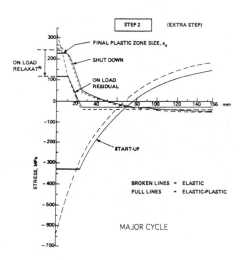

Fig. 7.8 Plastic zone size calculation

Table 7.1: Hysteresis Loop Results

Service cycle	$\Delta\varepsilon_t$, %	σ_0, MPa
Major, $\Delta T \sim 200°C$	0.684	240
Intermediate, $\Delta T \sim 150°C$	0.566	209
Minor, $\Delta T \sim 100°C$	0.353	109

In this example, an extra step is required, namely the estimation of the surface plastic zone size after the stress profiles have shaken down to a steady state as discussed in Section 6. The case for major cycling is shown in Fig. 7.8. A summary of results for other cycle types is given in Table 7.2.

Table 7.2: Depth of Plastic Zone

Service cycle	$\Delta\varepsilon_t$, %	Zone size, mm $(= a_o)$
Major	0.684	9.5
Intermediate	0.566	7.5
Minor	0.353	5.0

Note: This is another example of overlapping "zones within zones" according to the cycle mix, see Section 6.

7.3.3 Step 3

In order to calculate creep-fatigue damage, a continuous-cycling endurance plot was required. Clearly, to have determined the complete curve over all strain ranges would have required prolonged experimental effort. Instead, an established curve for a similar material was used and a single endurance obtained for the valve material at the calculated strain range of the major cycle. The result is shown in Fig. 7.9. Similarly, a ductility-strain rate curve was required for summing the creep damage as discussed in Section 3. An example for two extreme heat treatments of the chest material [5] is given in Fig. 7.10.

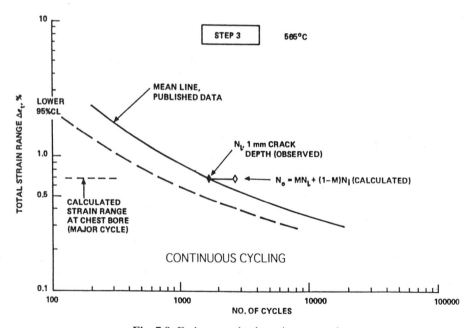

Fig. 7.9 Endurance check against mean data

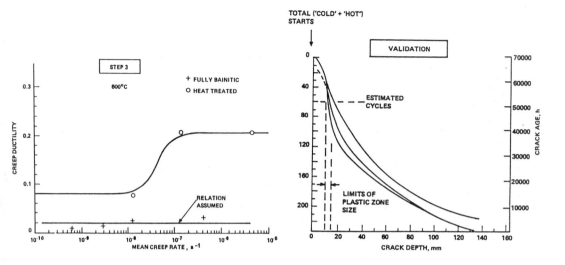

Fig. 7.10 Ductility transition curves   **Fig. 7.11** Oxide dating calculation

7.3.4 Step 4

As discussed in Section 3, the creep-fatigue cycles to fail the zone is given by:

$$N_o^* = \left(\frac{1}{N_o} + D_c \right)^{-1} \qquad\qquad .....(7.1)$$

In this case, however , we note that the maximum plastic zone size is somewhat *greater* than the diameter of a typical LCF specimen (6-8 mm). Since $Q = 1$ for the present material, referring to equation (4.4) it may be shown that the original terms N_o, a_o now become N_1, a_1 and replacing the original terms N_1, a_1 by the terms N_c, a_c (suffix 'c' for component) then:

$$N_c = \frac{N_1 - (1 - M)N_i}{M} \qquad\qquad .....(7.2a)$$

where:

$$M = \left\{ \frac{(a_{min} - a_i) + a_{min} \ln(a_1 / a_{min})}{(a_{min} - a_i) + a_{min} \ln(a_c / a_{min})} \right\} \qquad\qquad .....(7.2b) \cdot$$

The maximum value of plastic zone size (a_c) in this example is 9.5 mm.

The results of the damage calculation (i.e. cycles to fail the maximum zone size) are given in Table 7.3.

Table 7.3: Results of Damage Calculation
(Cycles to Fail 9.5 mm Plastic Zone Size)

Service Cycle	$\Delta\varepsilon_t$, %	N_o	$1/N_o$	Creep damage, D_c	N_o^*, predicted
Major	0.684	2712	0.00037	0.039	26
Intermediate	0.566	4339	0.00023	0.027	37
Minor	0.353	14,290	0.000069	< 0.002	483

One method of mixing cycles for a damage calculation is based on the time that the set is off load. In this case, only *major* and *minor* cycles are involved. Using the data in Table 7.3 therefore and assuming an 80% major ("cold start") and 20% minor ("hot start") distribution we have to solve the linear damage equation:

$$\frac{0.8N}{26} + \frac{0.2N}{483} = 1 \qquad\qquad .....(7.3)$$

giving $N = 32$ cycles. Thus a 9.5 mm crack initiates in the plastic zone by 32 service start-ups.

7.3.5 Step 5 (Optional)

This step is presented last of all in order to compare alternative calculations.

7.3.6 Step 6

This step deals with a sensitivity analysis. Another method of mixing cycles for a damage calculation is based on the 'temperature differential' distribution of data, see Fig. 7.2. In this case, *major*, *intermediate* and *minor* cycles are involved. Using the data in Table 7.3 therefore and assuming a 19% major ("cold start"), 48% intermediate ("warm start") and 33% minor ("hot start") distribution we have to solve the linear damage equation:

$$\frac{0.19N}{26} + \frac{0.48N}{37} + \frac{0.33N}{483} = 1 \qquad\qquad .....(7.4)$$

giving $N = 48$ cycles. Thus a 9.5 mm crack initiates in the plastic zone by 48 service start-ups.

Using oxide dating techniques [71] it was established that out of the total of 233 service cycles that actually propagated the crack across the whole wall section, approximately 60 cycles were required to attain the edge of the calculated plastic zone, see Fig. 7.11. Both predictions (32 or 48) cycles are thus considered safe since they are less than the observed cycle number.

Note: A sensitivity analysis with respect to materials data was not carried out in the present instance, which has used best-estimate data. Changes in one parameter may promote or demote the effects of others. For example, a lower bound yield stress could cause an increase in strain and maximise fatigue damage but since the stress at the beginning of dwell is reduced, this may reduce the creep damage. Conversely, an upper bound yield assumption would tend to decrease the fatigue damage and increase that of the creep component.

7.3.7 Step 7

It is of course essential to report the results of an analysis so that an independent assessment is possible. The present problem has not been sensitive to the choice of continuous cycling fatigue data because the problem is creep-dominated. The example has demonstrated the 'failure avoidance' capability of Volume 3 of the R5 procedure [8]. It may also be noted that the example has used *service-exposed* material properties and that this supporting materials data (continuous cycling endurance, stress relaxation and cyclic stress-strain) can be obtained relatively quickly. In the present example, the greatest sensitivity lies in the choice of service cycle mix, underlying the importance of temperature and strain monitoring in operation.

A deep crack growth analysis (i.e. for penetration beyond the surface plastic zone) has also been carried out, similar to the example discussed in Section 6. The crack growth relation employed for service-exposed material was:

$$\frac{da}{dN} = 4.7 \times 10^{-7} \Delta K_{eff}^{3.6}$$

.....(7.5)

for growth expressed in mm/cycle.

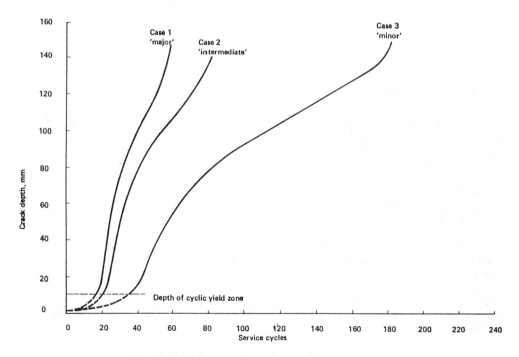

Fig. 7.12a   Integrated crack growth (separate)

In Fig. 7.12a is shown the penetration with cycles calculated for each cycle type acting separately. The more rapid advance with major cycles is clearly demonstrated. To check whether crack advance is also sensitive to the choice of cycle mix, several instances were tried as shown in Fig. 7.12b. It is seen that the overall crack growth rate is indeed sensitive to the mixture of cycles adopted.

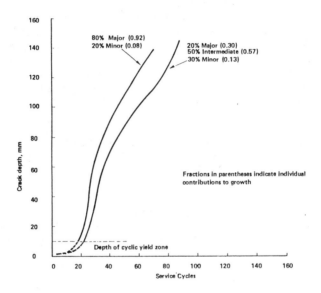

Fig. 7.12b   Integrated crack growth (mixed)

7.4. Damage Diagram

We end with plotting the present crack initiation results on the Damage Diagram (Step 5) in the manner discussed in Section 3. Referring to equations (7.3) and (7.4) and Table 7.3 and assuming a calculated 60 service cycles for the crack to traverse the plastic zone, in Fig. 7.13 points are plotted on the Diagram for:

- Major cycles only
- A mix of 80% major and 20% minor cycles
- A mix of 19% major, 48% intermediate and 33% minor cycles

These are listed in decreasing order of severity but in each case they lie in the "check for cracking" part of the Diagram i.e. the calculations on a linear damage theory have been validated.

Owing to the nature of the failure, the results are heavily biased towards creep rather than fatigue damage. It would be profitable to seek examples of the Diagram where fatigue is the dominant mode.

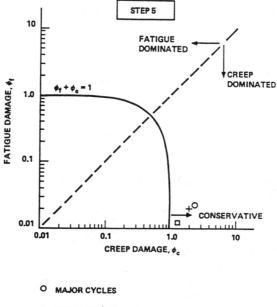

Fig. 7.13 Damage Diagram (for 60 cycles)

ACKNOWLEDGEMENTS

The author wishes to thank the successor companies PowerGen plc, National Power plc and Nuclear Electric Ltd for permission to publish that part of the work carried out under the auspices of the former CEGB.

REFERENCES

1. Skelton, R. P.: Introduction to thermal shock, *High Temp. Technol.*, **8** (1990), 75-97.
2. Skelton, R. P.: Application of small specimen crack growth data to engineering components at high temperatures: A review, in: *Low Cycle Fatigue* (Ed. H. Solomon et al.), ASTM STP 942, Philadelphia 1988, 209-235.
3. Skelton, R. P.: Fatigue crack growth, in: *Characterisation of High Temperature Materials: Mechanical Testing*, Institute of Metals, London 1988, 108-172.
4. Levaillant, C. and A. Pineau: Assessment of high temperature low cycle fatigue life of austenitic stainless steels by using intergranular damage as a correlating parameter, in: *Low Cycle Fatigue and Life Prediction* (ed. C. Amzallag et al.), ASTM STP 770, Philadelphia 1982, 169-193.

5.   Rémy, L. and R. P. Skelton: Damage assessment of components experiencing thermal transients, in: *High Temperature Structural Design*, ESIS 12 (Ed. L. H. Larsson), Mechanical Engineering Publications, London 1992, 283-315.

6.   ASME Boiler and Pressure Vessel Code, Case N-47 (29), Class I components in elevated temperature service, Section III, Division I, ASME, New York 1991.

7.   RCC-MR: Technical Appendix A3, Section 1, Subsection Z, Materials design and construction rules for mechanical components of FBR nuclear test islands, AFCEN, Paris 1985.

8.   Ainsworth, R. A. (Ed.): Assessment procedure for the high temperature response of structures, Issue 2, Nuclear Electric Ltd., Barnwood UK 1996.

9.   Skelton, R. P.: Developments in creep-fatigue crack initiation and growth procedures in high temperature codes, in: *Mechanical Behaviour of Materials at High Temperatures* (Ed. C. M. Branco), Kluwer Academic Publications, Dordrecht 1996, 281-297.

10.   Yoshida, S.(Ed.): Elevated temperature fatigue properties of engineering materials, Trans NRIM, Tokyo 1977, Vol. 19 onwards.

11.   Conway, J. B, Stentz, R. H. and J. T. Berling: *Fatigue, Tensile and Relaxation Behaviour of Stainless Steels*, US Atomic Energy Commission, Ohio 1981.

12.   Skelton, R. P. and G. A. Webster: History effects on the cyclic stress-strain response of a polycrystalline and single crystal nickel-base superalloy, *Mater. Sci. Eng.*, **A216** (1996), 139-154.

13.   Masuyama, F., Setoguchi, K., Haneda, H. and F. Nanjo: Findings on creep-fatigue damage in pressure parts of long-term service-exposed thermal power plants, Paper No. PVP-MF-84-015, ASME, New York 1985.

14.   Miller, D. A. and R. H. Priest: Materials response to thermal-mechanical strain cycling, in: *High Temperature Fatigue: Properties and Prediction* (Ed. R. P. Skelton), Elsevier Applied Science, London 1987, 113-175.

15.   Skelton, R. P. and L. Miles: Crack propagation in thick cylinders of 1/2CrMoV steel during thermal shock, *High Temp. Technol.*, **2** (1984), 23-34.

16.   Skelton, R. P. and K. Nix: Crack growth behaviour in austenitic and ferritic steels during thermal quenching from 550°C, *High Temp. Technol.*, **5** (1987), 3-12.

17.   Biot, M. A.: New methods in heat flow analysis with application to flight structures, *J. Aeronaut. Sci.*, **24** (1967), 857-873.

18.   Manson, S. S.: *Thermal Stress and Low Cycle Fatigue*, McGraw Hill, New York 1966.

19.   Houtman, J. L.: Inelastic strains from thermal shock, *Machine Design*, March 1974, 190-194.

20.   Simpson, I. C., private communication.

21.   Rees, C. J.: Thermal fatigue properties of candidate materials for replacement superheater headers, in: *Steam Plant for the 1990's*, Paper C386/025, Inst. Mech. Engrs., London 1990, 161-168.

22.   Skelton, R. P. and B. J. E. Beckett: Thermal fatigue properties of candidate materials for advanced steam plant, in: *Advances in Material Technology for Fossil Power Plants*, ASM, Ohio 1987, 359-366.

23.   Quentin, G. and W. Perez-Daple: Start-up control changes offer extended life for combustion turbines, *Controls and Automation Update*, EPRI, Palo Alto 1989, 2-3.

24.   Spencer, R. C. and D. P. Timo: Starting and loading of large steam turbines, in: *Proc. American Power Conf.*, Vol. 36, Chicago 1974, 511-521.

25.   Boller, C. and T. Seeger: *Materials Data for Cyclic Loading*, Parts A-E, Elsevier Science Publishers, Amsterdam 1987.

26. Bruhns, O. T. and H. Hübel: Rigorous inelastic analysis methods, in: *High Temperature Structural Design*, ESIS 12 (Ed. L. H. Larsson), Mechanical Engineering Publications, London 1992, 181-200.

27. Conle, A., Oxland, T. R., and T. H. Topper: Computer-based prediction of cyclic deformation and fatigue behaviour, in: *Low Cycle Fatigue* (Ed. H. Solomon et al.), ASTM STP 942, Philadelphia 1988, 1218-1236.

28. Skelton, R. P.: The relation between laboratory specimen and the practical case, in: *High Temperature Fatigue: Properties and Prediction*, Elsevier Applied Science, London 1987, 301-319.

29. Ramberg, W. and W. R. Osgood: Description of stress-strain curves by three parameters, Tech. Note No. 902, NACA 1943.

30. Skelton, R. P: Cyclic stress-strain properties during high strain fatigue, in: *High Temperature Fatigue: Properties and Prediction*, Elsevier Applied Science, London 1987, 27-112.

31. Rees, C. J. Skelton, R. P. and E. Metcalfe: Materials comparisons between NF616, HCM12A and TB12M - II: Thermal fatigue properties, in: *New Steels for Advanced Plant up to 620°C* (Ed. E. Metcalfe), EPRI/National Power plc, Swindon 1995, 135-151.

32. Timo, D. P.: Designing turbine components for low-cycle fatigue, in: *Thermal Stresses and Thermal Fatigue* (Ed. D. J. Littler), Butterworths, London 1971, 453-469.

33. Neuber, H.: Theory of stress concentration for shear-strained prismatical bodies with arbitrary non-linear stress-strain law, *Trans. ASME, Ser. E*, **28** (1961), 544-550.

34. Sumner, G. and V. B. Livesey (Eds): *Techniques for High Temperature Fatigue Testing*, Elsevier Applied Science Publishers, London 1985.

35. Skelton, R. P.: Energy criterion for high temperature low cycle fatigue failure, *Mater. Sci. Technol.*, 7 (1991), 427-439.

36. Coffin, L. F.: A study of the effects of cyclic thermal stresses on a ductile metal, Trans. ASME Ser. A, 76 (1954), 931-950.

37. Batte, A. D.: Creep-fatigue life predictions, in: *Fatigue at High Temperature* (Ed. R. P. Skelton), Applied Science Publishers, London 1983, 365-401.

38. Priest, R. H. and E. G. Ellison: An assessment of life analysis techniques for fatigue-creep situations, *Res Mech.*, 4 (1982), 127-150.

39. Thomas, G. and R. A. T. Dawson: The effect of dwell period and cycle type on high strain fatigue properties of a 1CrMoV rotor forging at 500-550°C, in *Engineering Aspects of Creep*, Vol. 1, Paper C335/80, Inst. Mech. Engrs., London 1980

40. Viswanathan, R.: *Damage Mechanisms and Life Assessments of High Temperature Components*, ASM International, Ohio 1989.

41. Skelton, R. P., Beech, S. M., Holdsworth, S. R., Neate, G. J., Miller, D. A., and R. H. Priest: Round robin tests on creep-fatigue crack growth in a ferritic steel at 550°C, in: *Behaviour of Defects at High Temperatures* (Eds R. A. Ainsworth and R. P. Skelton), ESIS 15, Mechanical Engineering Publications, London 1993, 299-325.

42. Ostergren, W. J.: A damage function and associated failure equations for predicting hold time and frequency effects in elevated temperature low cycle fatigue, *J. Testing Eval.*, **4** (1976), 327-339.

43. Ainsworth, R. A.: Defect assessment procedures at high temperature, *Proc. SMIRT 10 Conf.*, Anaheim, Ca., Vol. L (1989), 79-90.

44.   Feltham, P.: Stress relaxation in copper and alpha brasses at low temperatures, *J. Inst. Metals*, **89** (1960), 210-214.

45.   Batte, A. D., Murphy, M. C. and M. B. Stringer: High-strain fatigue properties of a 0.5CrMoV turbine casing steel, *Metal. Technol.*, 5 (1978), 405-413.

46.   Robinson, E. L. :Effect of temperature variation on the long time rupture strength of steels, *Trans. ASME*, 74 (1952), 777-781.

47.   Priest, R. H. and E. G. Ellison: A combined deformation map-ductility exhaustion approach to creep-fatigue analysis, Mater. Sci. Eng., 49, (1981), 7-17.

48.   Palmgren, A.: Die lebensdaur von kugellagern, Z. *Vereines Deutscher Ing.*, **68** (1924), 339-341.

49.   Miner, M. A.: Cumulative damage in fatigue, *J. Appl. Mech.*, **12** (1945), A159-A164.

50.   Skelton, R. P.: High strain fatigue testing at elevated temperature: A review, *High Temp. Technol.*, **3** (1985), 179-194.

51.   Van Den Avyle, J. A.: Low cycle fatigue of tubular specimens, *Scripta Metall.*, 17 (1983), 737-740.

52.   Pineau, A.: High temperature fatigue behaviour of engineering materials in relation to microstructure, in: *Fatigue at High Temperature* (Ed. R. P. Skelton), Applied Science Publishers, London 1983, 305-364.

53.   Skelton, R. P.: Growth of short cracks during high strain fatigue and thermal cycling, in: *Low Cycle Fatigue and Life Prediction* (ed. C. Amzallag et al.), ASTM STP 770, Philadelphia 1982, 337-381.

54.   Smith, D. J.: The behaviour of short cracks at elevated temperatures, *Mechanical Behaviour of Materials at High Temperatures* (Ed. C. M. Branco), Kluwer Academic Publications, Dordrecht 1996, 195-215.

55.   Hales, R. and R. A. Ainsworth: Multiaxial creep-fatigue rules, *Nucl. Eng. Design*, **153** (1995), 257-264.

56.   Kandil, F. A. and B. F. Dyson: Uncertainties in uniaxial low cycle fatigue measurements due to load misalignment, in: *Materials Metrology and Standards for Structural Performance* (ed. B. F. Dyson et al.), Chapman & Hall, London 1995, 134-149.

57.   Wareing, J., Tomkins, B. and I. Bretherton: Life prediction in austenitic stainless steel, in: *Flow and Fracture at Elevated Temperatures,* ASM, Ohio 1985, 251-278.

58.   Argo, H. C., DeLong, J. F., Kadoya, Y., Nakamura, M. and K. Ando, Eddystone experience on long-term exposed 316ss steam turbine valve components, ASME Paper 84-JPGC-Pwr-15, New York 1984, 1-11.

59.   Kimura, K., Fujiyama, K. and M. Muramatsu, Creep and fatigue life prediction based on the non-destructive assessment of material degradation for steam turbine rotors, in: High *Temperature Creep-fatigue* (Eds R. Ohtani et al.), Elsevier Applied Science, London 1988, 247-270.

60.   O'Connor, D. G. and J. M. Corum: Design rule for fatigue of welded joints in elevated-temperature nuclear components, in: *Symposium on ASME Codes and Recent Advances in PVP and Valve Technology Including a Survey of Operations Research methods in Engineering*, PVP-Vol. 109, ASME, New York 1986, 69-75.

61.   F. Engel, private communication.

62.   Maiya, P. S.: Considerations of crack initiation and crack propagation in low-cycle fatigue, *Scripta Metall.*, 9 (1975), 1141-1146.

63.   Manson, S. S. and M. H. Hirschberg: Crack initiation and propagation in notched fatigue specimens, in: *Proc. 1st Conf. on Fracture* (Ed. T. Yokobori et al.), Vol. 1, Japanese Society for Strength and Fracture of Materials, Sendai 1966, 479-499.

64.   Raynor, D. and R. P. Skelton: The onset of cracking and failure criteria in high strain fatigue, in: *Techniques for High Temperature Fatigue Testing* (Eds G. Sumner and V. B. Livesey), Elsevier Applied Science Publishers, London 1985, 143-166.65.

65   Skelton, R. P.: Damage factors during high temperature fatigue crack growth, in: *Behaviour of Defects at High Temperatures* (Eds R. A. Ainsworth and R. P. Skelton), ESIS 15, Mechanical Engineering Publications, London 1993, 191-218.

66.   Skelton, R. P.: Environmental crack growth in a 0.5CrMoV steel during isothermal high strain fatigue and temperature cycling, *Mater. Sci. Eng.*, **35** (1978), 287-298.

67.   Tomkins, B.: Fatigue crack propagation in metals: An analysis, *Phil. Mag.*, **18** (1968), 1041-1066.

68.   Ohtani, R. and T. Kitamura: Creep-fatigue interaction under high temperature conditions, in: *Crack Propagation in Metallic Structures* (Ed. A. Carpinteri), Vol. 2, Elsevier, Amsterdam 1994, 1347-1383.

69.   Miller, K. J.: Materials science perspective of metal fatigue resistance, *Mater. Sci. Technol.*, **9** (1993), 453-462.

70.   Priest, R. H., Miller, D. A., Gladwin, D. H and J. Maguire: The creep-fatigue crack growth behaviour of a 1CrMoV rotor steel, in: *Fossil Power Plant Rehabilitation*, ASM, Ohio 1989, 31-37.

71.   Pinder, L.: Oxide characterisation for service failure investigations, *Corr. Sci.*, **21** (1981), 749-763.

72.   Skelton, R. P. and J. Byrne: Prediction of frequency effect in high temperature fatigue crack growth using damage factors, *Mater. at High Temp.*, **12** (1994), 67-74.

73.   Levaillant, C., Grattier, J., Mottot, M. and Pineau: Creep and creep-fatigue intergranular damage in austenitic stainless steels: discussion of the creep-dominated regime, in: *Low Cycle Fatigue* (ed. H. D. Solomon et al.), ASTM STP 942, Philadelphia 1982, 414-437.

74.   Paris, P. C. and F. Erdogan, A critical analysis of fatigue crack propagation laws, *J. Basic Eng. (Trans. ASME)*, **85** (1963), 528-534.

75.   Haigh, J. R. and R. P. Skelton, A strain intensity approach to high temperature fatigue crack growth and failure, *Mater. Sci. Eng.*, **36** (1978), 133-137.

76.   Dowling, N. E.: Crack growth during low cycle fatigue of smooth axial specimens, in: *Cyclic Stress-strain and Plastic Deformation Aspects of Crack Growth*, ASTM STP 637, Philadelphia 1977, 97-121.

77.   Starkey, M. S. and R. P. Skelton: A comparison of the strain intensity and cyclic *J* approaches to crack growth, *Fatigue Eng. Mater. Struct.*, **5** (1982), 329-341.

78.   Shih, C. F. and J. W. Hutchinson: Fully plastic solutions and large scale yielding estimates for plane stress crack problems, *J. Eng. Technol.(Trans. ASME Ser. H)*, **98** (1976), 289-295.

79.   Athanassiadis, A., Boissenot, J. M., Brevet, P., Francois, D. and A. Raharinaivo: Linear elastic fracture mechanics computations of cracked cylindrical tensioned bodies, *Int. J. Fract.*, **17** (1981), 553-566.

80.   Dowling, N. E.: Geometry effects and the *J* integral approach to elastic-plastic fatigue crack growth, in: *Cracks and Fracture*, ASTM STP 601, Philadelphia 1976, 19-32.

81.   Skelton, R. P.: Cyclic crack growth and closure effects in low alloy ferritic steels during creep-fatigue at 550°C, *High Temp. Technol.*, **7** (1989), 115-128.

82.   Skelton, R. P., Priest, R. H., Miller, D. A. and C. J. Rees: Validation and background of crack opening and closing relation for use in high temperature assessment, *Fatigue Fract. Eng. Mater. Struct.* to be published.

83.   Skelton, R. P. and K. D. Challenger: Fatigue crack growth in 21/4Cr1Mo steel at 525°C II: Prediction of continuous cycling endurances, *Mater. Sci. Eng.*, **65** (1984), 283-288.

84.   Skelton, R. P.: Crack growth and cyclic stress-strain properties of 9Cr1Mo steel at elevated temperature, in: *Fatigue à Haute Température*, Societé Française de Metallurgie, Paris 1986, 185-203.

85.   Skelton, R. P. and J. I. Bucklow: Cyclic oxidation and crack growth during high strain fatigue of low alloy steel, *Metal. Sci.*, **12** (1978), 64-70.

86.   Buchalet, C. B. and W. H. Bamford, Stress intensity factor solutions for continuous surface flaws in reactor pressure vessels, in: *Mechanics of Crack Growth*, ASTM STP 590, Philadelphia (1976), 385-402.

87.   Hasan, S. T. and M. W. Brown: Thermal downshock fatigue testing in AISI 316 stainless steel plate, *High Temp. Technol.*, **2** (1984), 89-97.

BASIC MECHANISMS OF CREEP AND THE TESTING METHODS

E. Czoboly
Technical University of Budapest, Budapest, Hungary

ABSTRACT

Pointing at the importance of material selection in design process, the responses of structural materials to loading are discussed. The response can be an elastic or a plastic deformation or the material can fracture. These physical processes are discussed briefly examining them in atomic scale as well as in microscopic and in macroscopic sense. The influence of loading conditions as e.g. temperature or alternating loading are discuss too. The damage processes due to the loading are also shown.

The usual testing methods are overviewed and the different material characteristics are criticised. It is shown that although the testing methods are simple models of the real loading conditions, the most material parameters are not well defined indicators and therefore their misuse can result serious mistakes. Difficulties in fatigue and creep testings are exposed.

1 INTRODUCTION

The purpose of engineering activity is to design, to produce and to operate machines and structures of different kind. The working conditions of these products are very miscellaneous considering both the loads and the environment. Some parts are subjected only to static loads, however the most of them are loaded dynamically or cyclically. Even those structures, which are regarded as "statically loaded" ones have some variation in forces or other loading conditions. The environment of the structures may also be different. Temperature is only one, but important parameter of service. It can be high or low comparative to the ambient temperature and in general it is not constant during the service life. The rate of variation is also an important factor, mainly in the case of big structures. Rapid changes of temperature generate thermal stresses, which have to be regarded. Another deciding parameter is the surrounding media, which can be neutral or corrosive.

Some structures are operated under extremely hard conditions, so e.g. in that case, when the unfavourable factors are combined, the acting forces and the temperature vary frequently and in a wide range, corrosive media is present, etc. However, inspite of all these circumstances, we expect a perfect operation of the objects at least for a given period. The operation must be safe, therefore the life prediction procedures have to be improved continuously.

2. SELECTION OF MATERIAL

The behaviour of the structures depends very much on the materials they are made of. The selection of a proper material for a given machine part is a desicive action of the engineer. Although it can happen that for some special application there exist only very few or only one particular suitable material, for most purposes a variety of materials can be used. To find a satisfactory choice is not easy, it is necessary to consider the matter from different points of view.

First of all the mechanical, physical and chemical properties of the materials have to be taken into account in regard of the requirements. Here, of course, not only the service temperature and the acting stresses should be considered, but also the variation of these parameters, the frequency and number of cycling, the corrosive media, if any and the possible protection against corrosion, the probability of wear, erosion, etc. An important factor, which has to be regarded is the amount of risk of failure. The requirements are much higher, if the breakdown of the given part can result the loss of human life, or if it has a consequence of great financial damage.

Another factor of consideration in the procedure of material selection is the technological aspect. The possibly applied technology for production depends not only on the shape of the part, but also on the selected material itself and not less on the facilities of the given factory, where the part or structure will be produced. In this respect beside the available technical equipment, the effect of the technological process on the properties of the material (cold working, welding, etc.) and the skill of manpower should be regarded too.

Not less important is the economical aspect. Engineers are working for the market and the products have to be sold in a competition. Therefore, it is important that the possibly cheapest material should be selected together with an inexpensive but suitable technology. Sometimes - if not always - this has a decisive influence on the success of the product.

Beside the mentioned aspects some others may be important in given cases, such as esthetical or saniterial considerations. But these are apart from the topic of this course.

Although, the technological and economical aspects are very important, in the following only the material properties will be discussed, because the other mentioned requirements are strongly dependent on local circumstances variing from county to country and from factory to factory.

3. MATERIAL PARAMETERS

The properties of the materials can be characterized in two totally different ways: by words or by numbers. It is well known that in standards or catalogues such remarks can be found, as *weldable, resistant against atmospheric corrosion* and similars. Such qualifications are extremely important and they can help the experts a lot in their orientation, but it is obvious that they are not self evident and they need additional explanations.

On the other hand, some other properties of the materials are defined numerically. The numbers, called also *material parameters* are based on special experiments, generally known as *material testing methods*. These parameters are indispensable in dimensioning, therefore, they are extremely important in engineering practice. Of course, an accurate designing needs correct material parameters, which are based on real physical phenomena and are expressed in terms of realistic specific values. Unfortunately, these requirements are not fulfilled in the case of the majority of the used material characteristics.

At the present, we have only few parameters, which are correctly defined from a physical point of view and which can serve as a basis for dimensioning. The other parameters can be used only for guidance and for ranking the materials instead of exact calculations. Such parameters are in fact only arbitrary numbers. To make use of these conventional material characteristics former experiences are necessary; without such knowledge the information-content is very poor. Particular examples will be discussed later.

Of course, the general effort of material sciences is to increase the number of the "right" material parameters and to fill up the conventional ones with an improved physical meaning. The ellaboration of fracture mechanics is an outstanding example for this. New parameters very often need new testing methods, but a new interpretation of the conventional test results is also possible.

To understand the real meaning of the material parameters and treat them critically, a detailed analysis of the processes due to loading under different service conditions is necessary [1]. If a part is exposed to forces, it may respond in different ways. The most important response of the specimens regarding mechanical behaviour is *deformation*, which can be elastic or plastic. Elastic deformation is an inevitable consequence of the load, and can not be avoided. Plastic deformation may occur locally or in general. The former type of plastic deformation is in many constructions acceptable, while the latter should be avoided. Another consequence of the load can be *fracture* (partly or total), which is also an unwanted phenomenon and has to be avoided.

Because elastic deformation always occures, if the test piece is exposed to a load, the material testing methods try to determine those limit loads, which result the beginning of plastic deformation or fracture. The general problem is that the mentioned phenomena appear in very different combinations and their separation from each other is sometimes very difficult, if not impossible in the case of an industrial laboratory. In the following the mentioned processes will be discussed in details.

The analysis can be done on various levels dependent on the sizes of some basic units. In engineering material science generally three levels are used:

- atomic level, where the basic unit is an elementary cell. Forces among atoms (ions) and displacements of atoms are considered. Defects in the crystallic order as vacancies and dislocations, etc. are studied.

- microscopic level, where the basic unit is a grain. The role of different formations of grains, grain size, grain boundaries, various phases, precipitations, inclusions are discussed.

- macroscopic level, where the basic unit is the test piece or machine part. Acting forces, stress distributions, stress concentrations, strains, deformation, equilibrium, collapse and many other processes will be treated that way.

Although the test pieces used in the material testing methods represent usually the macroscopic level, but to understand what is going on within the material, an analysis on all levels is necessary.

4. DEFORMATIONS

4.1 Elastic Deformation.

Regarding the atoms in reality forces and displacements exist, but in-engineering practice the terms of stresses (σ or τ) and strains (δ or ε) are generally used. Stress is the force acting on a unit area

$$\sigma = F/A \qquad (4.1)$$

where F is load, A is area, Although, it is not very reasonable to talk about "stresses" on atomic level, because it is hard to define a cross section of an atom, this term is generally used, assuming that the same atomic forces are acting among many atoms, which form together a unit cross section.

The displacements are expressed as elongation or compression dependent on the relative direction of the motion. Finally, both are transformed to strains, which is a *specific elongation (or compression)*, that is the displacement of a unit of length:

$$\delta = (L-L_O)/L_O = \Delta L/L_O \qquad (4.2)$$

L_O is the original and L the extended length, ΔL is the elongation or the compression.

The elastic deformation is unavoidable, if any force is acting on a solid body. It is due to the interatomic forces, which are balancing the external load. *Fig.1.* shows the interatomic force of a simple arrangement as a function of the displacement. It can be seen that in the equilibrium position the attractive and repulsive forces are in balance, the resultant is zero. If the atom will be compressed to its neighbour, the repulsive force increases rapidly. If the interatomic distance expands, the growth of the attractive force is somewhat slower and reaches a maximum. However, if the displacements are very small, and this is so in the case of elastic strains, the relationship between stresses and strains (force and displacement) can be regarded, as linear.

The atoms of metals and other crystallic materials can be modelled by elastic balls, which are arranged in a geometrical order according to the *lattice*. In equilibrium state the balls are spherical, representing that the interatomic forces

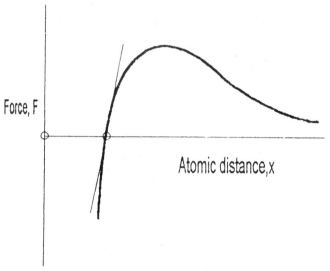

Figure 1. Interatomic force vs. displacement of atoms. In the case of small displacements the relation is nearly linear.

are balanced in any direction (*Fig.2/a.*). If an external load is acting on the crystal the balls in the lattice will be distorted. They may elongate in the direction of the load and contract in the lateral directions (*Fig.2/b*), or they may be deformed by an angular rotation (*Fig.2/c*). In the case of a complex, multiaxial loading the principle of superposition is valid, so the stresses and strains are additive.

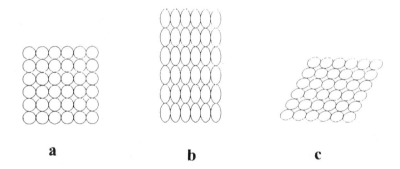

a   b   c

Figure 2. Arrangement of atoms in a simple cubic lattice. a) No acting stresses. b) Tensile stresses. c) Shear streses.

The general description of the deformation - considering a complicated atomic structure and a three-axial stress state - needs 36 elastic constants [2]. Fortunately, because of symmetry in the atomic arrangements, the number of the constants is strongly reduced. The most frequently used structural metals have e.g. a relatively simple crystallographic lattice, most of them belong to the BCC or FCC, that is to one of the cubic systems. For cubic crystals only three constants are enough.

$$\sigma_{11} = C_{11}\varepsilon_{11} + C_{12}\varepsilon_{22} + C_{12}\varepsilon_{33}$$

$$\sigma_{22} = C_{12}\varepsilon_{11} + C_{11}\varepsilon_{22} + C_{12}\varepsilon_{33}$$

$$\sigma_{33} = C_{12}\varepsilon_{11} + C_{12}\varepsilon_{22} + C_{11}\varepsilon_{33} \qquad (4.3)$$

$$\tau_{23} = C_{44}\gamma_{23}$$

$$\tau_{13} = C_{44}\gamma_{13}$$

$$\tau_{12} = C_{44}\gamma_{12}$$

Because of the crystall geometry, the deformation will be dissimilar in the different crystallographic directions. The difference depends not only on the type of lattice, but also on the material itself. E.g. tungsten and iron are both BCC metals. But the elastic constants for tungsten are in the different crystallographic directions about the same, while for iron they vary in a ratio about 1:2.

In policrystalline materials the grains are generally randomly oriented. In such materials the various elastic constants equalize each other and so the macroscopic body behaves as an isotropic material. In such cases the number of the elastic constants reduces further and only two independent parameters remain. These are the Young's modulus, E and the Poisson's ratio, v. Two other parameters are also used, however, these can be expressed by the former ones:

The *bulk modulus, K* characterizes the change in volume, ΔV as a consequence of a hydrostatic stress, σ_h

$$\frac{\Delta V}{V} = \frac{\sigma_h}{K} \qquad (4.4)$$

where

$$K = \frac{E}{3(1-2v)} \qquad (4.5)$$

and the *shear modulus, G* connect the shear stresses with the angular distortion

$$\tau = G\gamma \qquad (4.6)$$

where

$$G = \frac{E}{2(1+v)} \qquad (4.7)$$

For plastics and other materials, which do not display linear elasticity, the relation is much more complicated.

One of the most important features of elastic deformation is that it is fully reversible after unloading. The relationship between stresses and strains is one of the basic and important properties characterising the mechanical behaviour of the materials. Furthermore, it is characteristic of this type of deformation that all atoms and all atomic bonds of the stressed volume take part in it. Theoretically, considering an isotropic material the strain observed on the macroscopic body is the same as of the individual atomic bonds. Because of the differences in the atomic forces acting in the various crystallographic directions, this is not realized in anisotropic materials. In this cases differences may exist in the strains and stresses from grain to grain and according to the different crystallographic directions.

Although the anisotropy of the elastic constants can be neglected in general by the engineers, it can be important in some cases. Cold rolling or some other technological processes can produce a crystallographic texture, when the majority of the grains tend to orient themselves in a similar direction. Under such circumstances the anisotropy may have a significance.

The temperature has only a slight influence on the elastic properties. Since the atomic spacings increase and the bonding forces decrease with increasing temperature, the elastic modulus decreases. However, the variation is relatively small and - even more important - the whole mechanism of elastic deformation remains unaltered [3].

4.2   Plastic Deformation

In contrary to the elastic deformation, plastic deformation is irreversible and can occur by various mechanisms. The most important mechanisms in metals are:

. glide of dislocations,

. twinning,

. stress directed diffusion,

. climbing of dislocation ,

. grain boundary sliding.

Theoretically all mechanisms can take place in a specimen at once. In practice, mainly the temperature and other circumstances determine, which of the mechanisms will dominate. At ambient temperature and in the case of plastic materials the most important mechanism is glide of dislocations. Twinning occurs mainly at very low temperature and/or if the material is brittle. The other three mechanisms are characteristic for creep, when the temperature is relatively high, $T > 0.5T_m$, where T_m is the melting temperature in Kelvin..

4.2.1. *Glide of Dislocations*

To discuss the details of dislocation theory is far beyond the scope of this paper. Here only some charateristics should be mentioned, which are important in respect of understanding the significance of the material parameters.   Such important feature is e.g. the strong localisation of the deformation process. It is well known, that during the deformation process series of parallel dislocations are moving on their slip planes (the *active* slip planes), while the other parts of the crystals remain unaltered. (*Fig.3.*). Even in strongly deformed crystals a few hundred parallel atomic planes - which are all potential slip planes according to their orientation - can be found between two adjacent, active slip planes and these regions are not influenced by the deformation. The shape of a deformed single crystal is therefore not a simply elongated modification of the original grain, but it is more like a sliced piece.

Figure   3.   Schematic illustration of a plastically deformed single crystal.

Another important feature follows from the fact that the slip planes are definite crystallographic planes, which determine the direction of slip. These directions are not parallel with the axes of principal stresses or strains, but incline to the axis of the principal stress even in the case of a single crystal (See *Fig.3.*). The plastic deformation will start in single crystals therefore at that load, when the component of the force parallel with the slip system reaches its critical value.

In a polycrystalline material the distribution of deformation among the grains is also very irregular, particularly at the beginning of yielding. Some of the grains already endure plastic deformation, while others are still elastically deformed. The glide of dislocations is also much more complicated, since the orientation of the individual grains are different. So the slip directions and the location of the active slip planes varies also from grain to grain. However, the grain boundaries have to be matched during and after the deformation process. It can be shown that in such cases at least five independent slip systems have to be activated that two adjacent grains should remain in contact without creating voids or cracks (*Fig.4.*).

It is easy to imagine that most of the activated slip systems are not favourably oriented (not inclined in nearly 45 degrees to the greatest principle stress) and so they need a much greater force, as the yield stress of a single crystal. That is the reason that polycrystalline materials have significantly higher yield points, than single crystals. Since the grain boundaries hinder the slip process, the strength of the material increases with decreasing grain size, because the ratio of grain boundaries to the bulk of grains

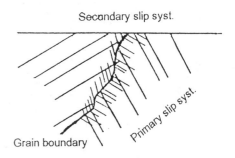

Figure 4. Multiple slip in a polycrystallin material

will be increased. This observation was formulated by Hall and Petch:

$$\sigma_Y = k_1 + k_2 d^{1/2} \tag{4.8}$$

where k_1 and k_2 are material constants.

Any technical metal and alloy have an original dislocation density of about $10^8/cm^2$. As the dislocations start to glide, dislocation sources develop and the number of dislocations increases rapidly. But the increasing dislocation density results also that they intersect each other frequently and the immobile jogs created this way hinder the motion of more and more dislocations. As a consequence, the number of mobile dislocations decreases and therefore the material displays strain hardening. This is illustrated schematically in *Fig. 5*. Similar to this phenomenon, any other mechanism, which hinders the motion of dislocations increases the yield strength of the material and reduces plasticity.

One of the generally applied methods to increase strength is to create dispersed, hard particals in the matrix. The size and density of these particals are deciding on their effectiveness. Other mechanisms, as refining the grains (increasing the amount of grain boundaries, as shown earlier) or introduce substitutional solute atoms into the crystal lattice are also effectual.

Temperature, strain rate and complexity of stress state influence the plasticity of the material strongly. Decreasing temperature and increasing strain rate generally decrease the mobility of dislocations or with other words, increase the critical stress, which is necessary to activate the slip systems. "Critical stress" means in this sense *shear* stress, because the activation of dislocation glide needs shear stress. Therefore, the greatest shear stress controls the process of plastic flow. In a hydrostatical stress state, when all the principal stress components have the same sign and magnitude, the shear stresses are equal to zero, so such kind of stress state - independent from the level of stresses - does not result plastic deformation. This can be proved by high pressures in liquids. In practice, a totally hydrostatical tensile stress state does not exist. However, the so called *plane strain* state is usual at a cracked, thick plate near the crack tip and its character is not far from the hydrostatical tensile stress state. Therefore, plasticity is in such cases strongly reduced.

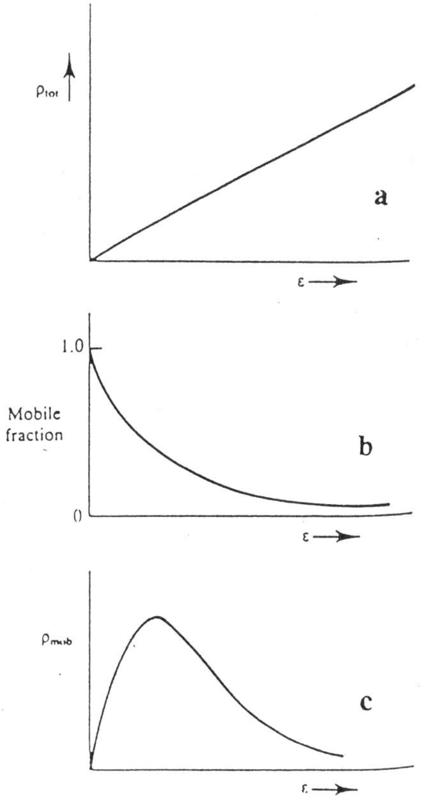

Regarding a test piece in a macroscopic manner the mentioned peculiarities are not considered. The deformation is measured globally for the whole test piece or at least in a restricted small region, which still includes a great number of grains, therefore the individual slip processes will contribute to a global strain, which is defined and calculated in a similar way, as in Eq.(4.2). However, one has to keep in mind that the elementary mechanism is *slip* resulting from *shear stress* and *not elongation.*

Figure 5. Schematic illustration of the a) total number of dislocations, b) fraction of mobile dislocations, c) number of mobile dislocations vs. strain [2}.

In contrast to elastic deformation, plastic strain is very much influenced by strain rate and temperature. The critical shear stress, which is necessary to move the dislocations and their mobility is a function of the parameters mentioned. The former determines the stress, which starts plastic deformation - called yield stress or yield point - the latter influences the amount of plastic deformation, which can be tolerated by a material without fracture. In general, increasing strain rate and decreasing temperature will increase the yield point and will decrease the plasticity of the material.

Temperature   effects also the behaviour of a specimen, if exposed to a constant or nearly constant load. The first reaction of the material to the load is a given amount of deformation, which is independent from time. This strain may have an elastic and also a permanent (plastic) component. Because of the strain hardening mechanism (decreasing number of mobile dislocations) this process stops very soon and a quasi equilibrium state develops. However, many dislocations are piled up at different obstacles, because the driving force is just below the necessary one to push them through. If the temperature is high enough, *thermal activation* can help the dislocations to pass through the particles and produce an additional, time dependent deformation. It is obvious that the higher is the temperature, the more mobil will be the dislocations, that is the greater will be the strain rate.

Finally, the effect of the deformation process on the internal stresses should be mentioned. While elastic deformation does not initiate residual stresses after unloading, plastic deformation leaves its marks. This is not only because of the deformation process itself, which results in an increase of the density of dislocations by some order of magnitude. The dislocations and the surrounding volumes mean a great disorder in the lattice, which can be considered as highly stressed zones. Beside this, due to the different directions of the active slip planes in the neighbouring grains, second order stresses remain. This has a decisive influence on the behaviour of the material, if alternating stresses are acting. Normally, after a plastic deformation the materials indicate strain hardening and the yield point will be raised. However, if the acting load will be reversed, the plastic flow starts at a much lower level, than the original yield strength. This phenomenon is known as Bauschinger effect.

4.2.2. *Twinning*

This mechanism of plastic deformation is only mentioned here for completeness, because it has a minor significance in connection with creep. Twinning deformation can be observed mainly in hexagonal materials. In BCC or FCC metals twinning occurs only at low temperatures and at high strain rates. The reason for this is that BCC and mainly FCC materials have enough slip planes, which can be activated. Slip needs less energy than twinning, therefore twins will nucleate only under special circumstances, when slip process is hindered.

The contribution of twinning to the deformation is small. Therefore those materials, which deform mainly by twinning, fracture after only a small amount of plasticity. In contrast to the elastic deformation, twinning - similar to slip process - is also strongly related to definite crystallographic planes and only a small fragment of the whole grain takes part in the process (*Fig. 6.*). As it can be seen in the figure, twinning means a kind of rotation of one part of the crystal mirrored to the twin plane of the original atomic arrangement.

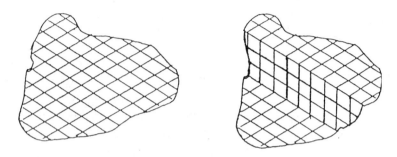

Figure 6. Twinning mechanism in a crystal.

Observing one individual grain, it is doubtful that the result of a twinning can be characterised by an "elongation" or "strain". These very common concepts have hardly any sense in the microscopic or in the atomic world. Of course, on a macroscopic scale one can measure global elongation or strain as a result of the simultaneous processes in many grains.

Elastic deformation and twinning will occure immediately, as the load is applied. Although, gliding of dislocations needs some time (hence its sensitivity for strain rate), the time is necessarily short, practically a few fragments of a second, so all the discussed mechanisms can be regarded as time independent reactions. If the load acts for a longer period, some more mechanisms beside the thermal activation mentioned earlier can take place and can contribute to the total deformation.

4.2.3. *Stress Directed Diffusion*

It is well known that the lattice contains always empty places, so called vacancies, the number of which is a function of temperature.

$$n/N = \exp(-\Delta U/RT) \qquad (4.9)$$

where n is the number of vacancies, N the number of the lattice points considered, ΔU is the activation energy of the vacancies, R is the gas-constant and T is the temperature in Kelvin. Near the melting point the n/N ratio reaches a value about 10^{-4}[3].

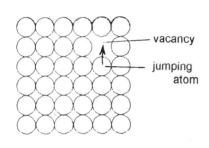

Figure 7. Basic mechanism of diffusion.

These vacancies are steadily wandering in the lattice, because the neighbouring atoms jump to the empty places. (*Fig. 7.*). If the atoms, which take part in this change-of-place process are from the matrix (atoms of the base metal), this steady movement is called *self-diffusion*. In contrast to the solution or segregation processes , when alloying atoms are wandering in the lattice, the self diffusion has no special direction or tendency. The motion of the atoms is random. However, if stresses are acting on the specimen, the motion of the atoms will get a definite trend: vacancies are mostly developing at those boundaries of the grains, which are perpendicular to the direction of the tensile stress and they are wandering to the parallel boundaries. On the other hand, atoms are migrating in the opposite direction (*Fig. 8*).

It is shown that at high temperatures and low stress levels this mechanism is controlling creep. It is generally called *Nabarro-Herring creep*. A very similar process can be obtained even at somewhat lower temperatures. Here, the diffusion occurs along the grain boundaries. This type of creep is called *Coble creep.*

From a comparison of the mechanisms discussed earlier, one can conclude that stress directed diffusion is in one particular way similar to elastic deformation. The similarity lies in the fact that the whole bulk of the grains or at least all grains take part in the action and the result is really an elongation even regarded on microscopic scale. The amount of possible deformation is in both cases small. Of course, other features of the two types of deformation are very different.

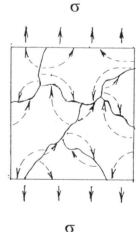

Figure 8. Migrating of atoms due to an active tensile stress. Vacancies move in the opposite directions.

4.2.4. Climbing of Dislocation

At higher stress levels even at moderate temperatures a new mechanism starts to be active. This can be regarded as the main process in creep, although in some special cases the contribution of other creep processes to the total deformation can be significant.

As mentioned earlier, after an initial loading and subsequent deformation, series of dislocations are piled up at different obstacles. *Fig.9/a* gives an example, where the motion of dislocations is prevented by a second phase particle. Driving force even together with a temporary thermal activation energy is not enough to help the dislocations to pass. However, with the aid of vacancies there is a possibility for them to avoid the obstacle.

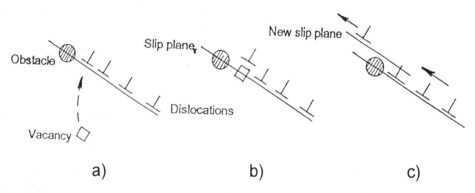

Figure 9. Non-conservative notion of dislocations by the help of migrating vacancies

In that temperature range, where creep can be expected $(T > 0,5T_m)$, the density of vacancies is relatively high. If a vacancy is migrating to the edge of the extra-plane, that is to the core of the dislocation, the slip plane of the dislocation will be raised by one atomic distance (*Fig. 9/b*). Of course, such kind of migrations are necessary over a longer range of the dislocation line and generally a slip plane has to be raised not only by one, but by a number of lattice spacings to get around the particle. If this happens, dislocations avoid barriers and can continue their motion on the new slip plane. So, plastic deformation will be continued. This kind of nonconservative motion of the dislocations is called *climbing (Fig.9/c)*.

The deformation of the grains according to the mechanism described above is the result of a combined motion, climbing and gliding of the dislocations. Climbing ensures the necessary number of mobile dislocations, which were formerly hindered by some barrier. Subsequent gliding produces plastic deformation on the new slip planes. Both processes are needed to continue deformation. Since they are differently influenced by temperature and by the acting stress, their possible rate will be also different. The resulting strain rate will be controlled by the slower process [2].

As for the other characteristics of this procedure, one have to keep in mind that the reason of the deformation is in fact gliding with all its tipical features. Therefore, all statments made in Chapter 4.2.1 concerning a comparison with other mechanisms are valid here too.

4.2.5. Grain Boundary Sliding

This mechanism does not occure alone in the materials, it can be regarded much more as a necessary consequence of the formerly mentioned other mechanisms. The situation illustrated in *Fig. 10.* is a tipical example. Imagine that the three grains shown in the figure are exposed to stress directed diffusion. Their dimensions will increase in the direction of tensile stress and will be reduced in the lateral directions. If the grains are supposed to keep together, the two upper grains have to slide on the lower grain to remain in contact with each other. In opposite case voids or cracks would be created.

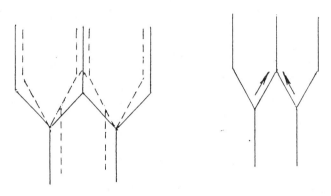

Figure 10. Constrained slip on grain boundaries to maintain the contact between the grains.

In real materials, however, the grain boundaries are not even planes, but they have a complex geometry. This is outlined in *Fig.11.* If such boundaries are going to slip on each other, voids and overlapping volumes are left behind. Overlapps are impossible. So, the redundant atoms have to leave their positions and move to the empty regions. Considering that the lattices in the grain boundary regions are anyhow disturbed, so the bonding energy is at a higher level (which means that the bonding energy is less, than in a perfect lattice), the atoms can be mobilized by a lower activation energy than in the bulk of the crystal.

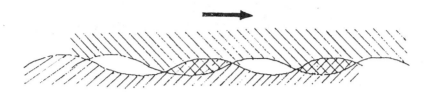

Figure 11. Nucleation of vacancies and overlapping areas as a consequence of grain boundary sliding.

The consequences of this mechanism are great, although the contribution to the whole deformation is small, therefore it is worth-while to examine it more closely. The disorder created by the boundary slip cannot be compensated completely by the diffusion mechanism. So, some voids and later even small cracks may remain. These cracks may join and propagate and even may lead to final fracture.

The smaller the grains and so the greater the grain boundary/ grain bulk ratio, the greater is the risk that imperfections nucleate. To reduce this unwanted effect of the grain boundaries, coarse grained materials have to be selected - the only case, when a coarse grained material is preferred because of its better mechanical properties in contrast to a fine grained material.

5.   FRACTURE

Fracture is an unwanted phenomenon in engineering practice and it should be avoided anyhow. In general it means that a machine part or structure is broken into two or more pieces. Under certain conditions, however, fracture can be defined also as the development of a crack of a given length, when the part has to be discarded.

Considering fracture process from the point of view of the atoms, it means the seperation of atoms or layer of atoms by breaking the atomic bonds. The force between two adjacent atomic layers as a function of the interatomic distance is shown in *Fig.12*. If no external force is acting, the spacing between them corresponds to the lattice parameter. Responding to external forces the distance increases and after exeeding a critical amount of elongation (see in the figure: Δ a_{crit}) the interatomic force approximate zero. It was demonstrated that the stress needed to such kind of separation, σ_{th} is equal to

$$\sigma_{th} = \sqrt{\frac{\gamma E}{a}} \qquad\qquad (5.1)$$

where γ is the surface energy, E the Young's modulus and a the lattice parameter [2]. At the same time it was shown that the materials never reach this theoretical strength, they fracture at a much lower stress level.

The example represented an absolute brittle, a so called cleavage fracture. In technical materials, mainly in metals such a brittleness is unusual. Metals always display more or less plasticity. Depending on the amount of plasticity preceding the separating process and the volume of the material involved into the plastic deformation, fractures can be classified into a brittle or into a ductile type. Although, this classification is self-evident in the extreme cases, in practice many transitional occurrences may happen, where the determination is often arbitrary and not obvious.

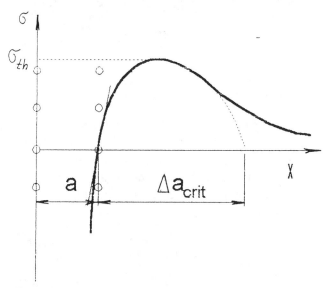

Figure 12. Stress between atomic layers as a function of displacement. Equilibrium distance is a; at a distance of $a+\Delta a_{crit}$ fracture is complete.

Fracture processes can be categorized also in a different way. The separation of the various parts may run through the grains or on the grain boundaries. The first type of fracture is called transgranular, while the second type intergranular fracture. The stresses, which result fracture may be considered also as the basis of classification. In most cases tensile stresses pull apart the two pieces to be fractured. However, it can happen that the parts are separated by shear. *Fig. 13.* gives examples for the mentioned fracture types. Once again it have to be emphasized that in the praxis the introduced types of fracture occure mixed [4].

Fracture behaviour of a material is in fact not a "material property", but it depends on the circumstances. The main influencing factors are temperature, strain rate and state of stress (uni- or multi-axiality of stresses and the sign of the stress components).

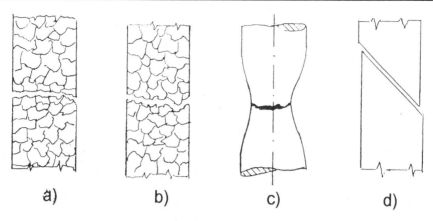

a)   b)   c)   d)

Figure 13. Different types of fracture. a) transgranular, b) intergranular, c) plastic, tensile fracture, d) shear fracture.

BCC metals are sensitive to temperature and transform their fracture behaviour from plastic to brittle as the temperature decreases. Similar curves, as seen in *Fig.14.* are well known. It is also a common knowledge that such

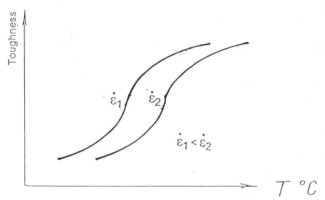

Figure 14. The effect of temperature and strain rate on the toughness of Bcc materials.

transitions have a very important role in engineering, but its significance is less in respect of creep. Creep is related to plastic deformation and in many cases great amount of plastic deformation precedes fracture indeed. However, this is not always the case. In some cases plasticity is strongly reduced in creep procedures and creep fracture is very brittle regarded macroscopically.

Strain rate is another important factor, which influences the type of fracture. The increase of strain rate has a similar effect, as decreasing temperature. Again, regarding creep, where very slow deformation processes occure, the problem of

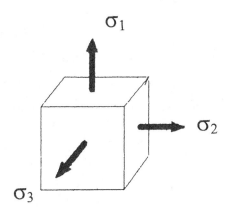

Figure 15. Dangerous multiaxial stress state. Worst, hydrostatical case: $\sigma_1=\sigma_2=\sigma_3$ Plain strain state: $\sigma_3=\nu(\sigma_1+\sigma_2)$

dynamic loadi ng can be disregarded, but other problems related to the very slow deformation rate arrise.

The third factor mentioned above, the stress state may have a significant effect on plasticity, even if other circumstances would favour plastic deformation, e.g. at relatively high temperature. Multiaxial stress state can support or hinder deformation processes dependent on the sign of the stress components. Three tensile components, as illustrated in *Fig. 15.* has a detrimental influence on any material. Since such a stress state is typical in the vicinity of a notch, crack or other defects, materials exhibit a strongly reduced plasticity at these places.

It is worth while to mention that some type of fractures induced by plastic deformation seem from an engineering point of view to be brittle, because the plastic deformation is restricted to a very narrow part of the material near the crack surfaces. Well known examples for such a behaviour are fatigue fractures and in some cases also creep ruptures. Fatigue process is strongly related to plastic deformation. Microcracks are induced by dislocation glide and propagation of macrocracks are also combined with a cyclic plastic deformation around the crack tip. However, a typical fatigue crack propagates through a part without any macroscopic deformation, the surrounding material does not reveal any plasticity.

This is less typical for creep rupture, but it may occure under specific circumstances that the crack propagates in a "quasi brittle" manner, although the process itself - regarding it from a microscopic point of view - is related to plastic deformation. Brittle fractures take place if intergranular fracture is preferred and frequently in agressive environments, where corrosive and cracking mechanisms are strengthening each other.

The most up-to-date theory, which deals and describes fracture processes: *fracture mechanics* [5] has demonstrated that fracture is not a single, but a dual process; it consists of a *crack nucleation* and a *crack propagation* period. Further it was proved that crack nucleation process can be neglected, because technical materials always contain a number of defects, which can be considered as potential crack nuclei. The deciding process is the onset of crack propagation and the most important question is, whether the crack will propagate in a stable or unstable manner.

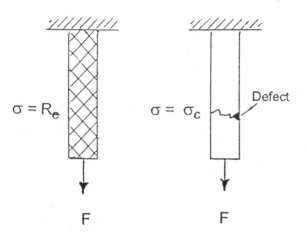

$$\sigma = R_e$$

$$\sigma = \sigma_c$$

Defect

F   F

Figure 16. Comparison of the behaviour of two specimens: a a plastic and a brittle one, pulled till the end of the elastic regime.

This observation means in other words that stresses alone are not responsible to result fracture; an adequate defect is also necessary. This can be demonstrated by a very simple example. Consider two tensile specimens (*Fig. 16*), one of them is made of a plastic and the other one of a brittle material. By increasing the load, the first specimen will reach the limit of elastic deformation and will start to deform plastically. At that moment the stress is uniformly distributed in the total volume of the specimen. Its value corresponds to the yield stress, R_e and as a consequence, yielding will take place in the whole specimen.

By increasing the load of the second specimen, a "critical" stress level, σ_c will be reached, which will generate fracture. But this physical process (fracture) is restricted only to a small fragment of the test piece, to a thin layer. In this thin layer a series of atomic bonds break, two atomic planes separate in each grain and probably some limited plastic deformation occures in the neighbouring atomic layers. Theoretically, all the other grains and atomic layers remained undisturbed, although the whole specimen was exposed to the same stress. The reason of this behaviour lies in the fact that the fractured cross section turned out to be weaker, than the others. There must have been a defect or with other words: *a crack,* even if this is unknown for the observer. Therefore, the value of σ_c is not a real material parameter, because it does not only depend on the material, but also on the size and geometry of the defect.

This statment was first formulated by Griffith [6], for an infinitely wide plate, with a through crack of a length $2c$. According to his formula

$$\sigma_c = \sqrt{\frac{2\gamma E}{c\pi}} \qquad\qquad [5.2]$$

It can be seen that the critical stress is dependent on the crack geometry, namely on c. Without going into details of fracture mechanics, it has to be remembered that the concepts of fracture mechanics are increasingly applied in the research of fatigue and creep [7]. It is more and more realised that the crack propagation period are extremely important in both processes. And fracture mechanics can be considered as the only reliable method today, which provides the possibility to model correctly the cracked structures by test pieces and so to transfer the results of material testing to the real structures.

On the other hand, fatigue and creep processes are strongly related to crack initiation too and therefore crack nucleation processes should be also thoroughly investigated. These research studies have their practical importance [8]. The weak point of these investigations is the definition of the evaluated crack, or otherwise the determination of the end of the crack initiation period and the onset of the crack propagation period.

Before a couple of years instead of existing, acceptable, theoretical definitions, very simple practical terms were used. According to recommendations in the literature, a 2 mm long crack was regarded as a macro crack, a so called "engineering crack" assuming that such a crack is visible by a nacked eye. This "definition" is still used, but now-a-days there are also more scientific ones. This will be discussed later.

6. MATERIAL TESTING

In the followings, some general statements should be made in relation to the peculiarities of material testing methods and their results. The analysis will be restricted to those testing methods, which aim to provide the engineers with *material parameters*. These are numbers, characterising the mechanical properties of the materials in general .

Of course, there are also other important methods, which cannot be missed in modern technical life. So, e.g. those tests, which do not destroy the tested parts and which aim the detection of included defects. Such NDT (non-destuctive test) methods are the ultrasonic test, the X ray method, magnetic and penetrant fluid test, etc. They are capable to indicate surface and embedded defects above a critical size. The limit of indication is an important factor in safety considerations, the prescribed testing methods and the interval of the repeated testing after a given service life are based on that. However, NDT methods are not subjects of this lecture.

Those procedures, which are testing ready components will be excluded also from this study. Here only those methods will be discussed, as mentioned above, which provide *numerical parameters*. The most important and mostly used tests of this kind are:

- tension test
- hardness measurement
- impact bending test
- fatigue tests
- creep tests
- fracture tests

A common feature of all these methods is that they are modelling the real loading processes, measuring the load and observing the reaction of the test piece. At loads of predefined events numerical parameters are constructed to characterise the material. Some of these numbers are correctly interpreted, taking into account the real physical processes, which happen during the testing or which just begin at the selected limiting load. These parameters have a real physical meaning and are strongly related to that physical phenomenon, what they characterize. Such parameters can be used in a wide range in the dimensioning process or in technology planning, even if the postulated circumstances are not exactly the same, as they were in the test. This means with other words that the results of the simplified, modelling loading processes (the tests) can be transferred to real conditions and structures.

Other parameters are often only numbers without physical meaning or with a more or less incorrectly constructed content. These can serve only for comparison, for material selection, ordering them in a sequence according to a special type of loading. Such parameters are not adequate for dimensioning. Design engineers can use such parameters only indirectly, supposing that enough practical experience has been accumulated.

In the followings at first some well known material charateristics are analysed for the sake of a better understanding of the physical meaning of these quantities. Afterwards those parameters will be examined, which are used in creep and fatigue and related processes.

Before going in details, however, it is necessary to emphasise again that even the best "material parameters" do not depend only on the material, but also on the environmental circumstances, so first of all on temperature, on strain rate and on stress state. One of the important tasks of material testing is just to determine these functions.

6.1. Tension Testing

The most commonly used test in engineering practice is the tension testing, or tensile test. It is also very adequate for a detailed analysis and for a demonstration, whether the calculated and widely used parameters are correct or incorrect.

The test is performed as follows: one end of a round or rectangular rod is fixed to a load cell, while the other end is pulled with a given, constant velocity. The specimen reacts to this stretching process with a resistance, which is measured by the load cell and registered on a graph: the so called *load-elongation diagram*. The shape of the diagram is representative for the materials and it is used to determine the important material parameters. For this purpose those loads are selected and measured, where the character of the diagram changes showing that a new process starts. For illustration the load-elongation diagram of a low carbon steel is the best, because every typical point and region can be observed on it. Such a diagram is shown in *Fig. 17*.

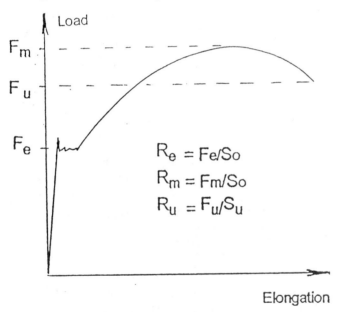

Figure 17. Typical tensile diagram of a mild steel.

The material parameters determined by the tension tests belong to two systems. One group of them belongs to the so called *engineering* parameters, where the reference dimensions are the original cross section and length of the specimens, while the other group consists of the *true* parameters, where the dimensions continuously varying during the tests are used, as references. The parameters can be distinguished also by their meaning. One group represents the strength of the materials, the other group the plasticity.

In the case of the engineering system the stress is calculated according to the original cross section, A_0 that is

$$\sigma = F/A_0 \tag{6.1}$$

where F is load. In the true system on the other hand, stress is

$$\sigma' = F/A \tag{6.2}$$

where A is the variing cross section. Similarly, engineering strain refers to the original length, L_O

$$\delta = \int dL/L_0 = (L-L_0)/L_0 = \Delta L/L_0 \tag{6.3}$$

and true strain to the variing, momentary length, L

$$\varepsilon = \int dL/L = \ln(L/L_0) \tag{6.4}$$

In case of small strains, e.g. elastic strains or around the yield stress the engineering stresses and strains equal to the true parameters. However, at greater strain values they deviate from each other and therefore, it is important to decide whether the *engineering* or the *true system* should be used.

The true stress/true strain curve, as shown in *Fig.18.* has much more realistic physical content, than the tensile diagram shown in *Fig.17.* However, in engineering practice the tensile diagram is more often used.

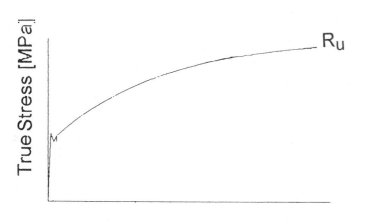

Figure 18. True stress/true strain curve

A tyical parameter, which is regarded as a physically well defined one and which is basically used for dimensioning is the *yield point,* R_e or σ_Y. It is related to the onset of macroscopic flow. It is assumed that untill this stress is reached the deformation is entirely elastic and the cross section of the test piece practically does not change. Therefore, R_e can be regarded equally as an engineering and also as a true stress. Although it is determined in a uniaxial stress state, the yield criteria can be calculated also in other, multiaxial stress states (see: yielding theories of Tresca or of von Mises), because it has realistic physical meaning (the onset of macroscopic flow).

However, even this "best" parameter can be criticized and it can lead to wrong conclusions, if some special circumstances are disregarded. In given state of affairs it has to be taken into consideration that although the yield point characterizes the onset of macroscopic flow, yielding may occur in some favourably oriented grains far below this stress level.

Furthermore, it should be remembered that only a few materials have a definite yield point as shown in the *Fig.17*. Most materials display a continuous transition from elastic to plastic behaviour. For these materials a $R_{p0,2}$ stress limit will be defined and used instead of the yield point. According to the definition this means that stress, where 0,2% permanent strain remains after unloading. It follows from the non-uniform distribution of the plastic deformation considered in a microscopic scale that significant plastic strain may occur in small regions even before this macroscopic limit will be reached.

The situation is even worse with the widely used *ultimate tensile stress, R_m* or *UTS,* which is the maximum force devided by the original cross section. The physical content of this parameter is very poor. The parameter is an engineering stress, but when the maximum load is reached, the cross section of the specimen had been reduced significantly (15-30% for usual structural materials). Therefore, the real value of the stress is by 15-30% higher at that stage of the test . Beside this, even its name is missleading, because R_m is *not* an *ultimate* stress. The *true fracture stress, R_u,* which acts at the end of the test, when fracture really occurs, is much higher (See *Figs.17 and 18.*). In contrary to all these imperfections R_m is frequently recommended even to dimensioning processes. However, in contrast to the yield strength, this value cannot be transformed to an equivalent stress in multiaxial stress states.

Ultimate tensile strength has been regarded in the past as one of the most important material parameters. Therefore, a number of correlations were established, as e.g. correlation with fatigue limit, etc. Such correlations are still mentioned in the literature, but because of the poor physical content of R_m, they have to be regarded cautiously.

The situation is not much better with the *strain*. In the standards an A_x average strain is defined. This parameter can be calculated from the elongation of the test piece measured on an arbitrarily selected (but in the standards prescribed) gage length, x. Since the local strain varies considerably along the test piece, reaching its maximum at the site of fracture, the "standard" strain is not independent from the gage length. In Europe *5d* or *10d* gage lenths are used, where d is the diameter of the specimen. In the English speaking countries other lengths are used. To convert these average strains into each other is an unreliable operation.

The standard strain values are only for comparison of the different materials. They cannot be used for any calculation. It is obvious therefore, that in research and in technology planning another definition of strain have to be applied. This is the *local strain*. Local strain is theoretically measured on an infinitely short gage length, practically it is calculated from the variation of the cross section of the specimen. If not indicated otherwise the local strain is always measured at the smallest cross section. Here also an engineering and a true systems can be distinguished. The *engineering local strain* can be defined as:

$$(L-L_0)/Lo = (L/L_0) - 1 = (A_0/A) - 1 \tag{6.5}$$

the *true local strain,* as

$$ln\ (L/L_0) = ln\ (A_0/A) \tag{6.6}$$

6.2. Absorbed Specific Energy Till Fracture

Both the concepts, *stress* and *strain* have the disadvantage of being vectorial quantities. Furthermore they are macroscopic concepts and cannot be interpreted in atomic scale. Energy is on the contrary a scalar quantity and it has sense either in atomic, or in macroscopic approach.

The concept of the *absorbed specific energy till fracture* (ASPEF) was recommended by Gillemot [9] and represents a material characteristic with a reasonable physical meaning. The most simple way to understand the concept of ASPEF and to measure it is a tension test. The basic idea is that both the plastic deformation preceding fracture, as the separation process itself need energy. This energy is supplied by the external load and is absorbed in the deformed part of the test piece. Since the deformation is not uniform in a necked tensile specimen, neither is the distribution of the absorbed energy.

Locally, the greatest energy will be absorbed in that part of the specimen, where the deformation reaches the highest value and where - afterwards - fracture occures. Expressing this energy as a specific value, a well defined material characteristic is constructed. The volume related to the absorbed energy is very small, it is only a thin slice enclosing the fracture surfaces.

So, ASPEF is defined, as the *specific* energy done by the external load and absorbed in that infinitely small *volume,* which contains also the fracture surfaces and which endured the greatest deformation (*Fig. 19.*). By this definition:

$$W_C = \int(F.dl)\ V \tag{6.7}$$

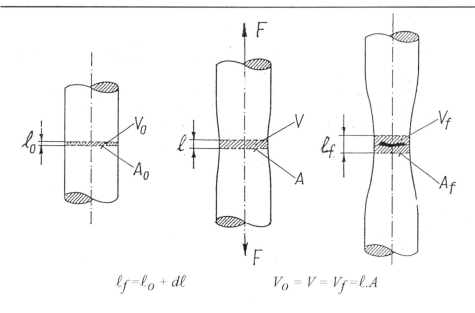

Figure 19. Illustration of the concept of Absorbed Specific Energy till Frecture, ASPEF.

Since the plastic deformation does not alter the specific volume of the material, V is constant during the whole test and because it is a very thin slice, the diabolic shape of the specimen can be disregarded and V can be simply described as $l.A$.

In Equ. (6.7) F is the varying load during the test, dl is the elongation of the infinitely small gage length, l_O selected at the site of the least cross section, V is the volume of the infinitely thin slice. Substituting $l.A$ instead of V, one obtains

$$W_C = \int (F/A).(dl/l) = \int \sigma'.d\varepsilon \qquad (6.8)$$

where σ' is true stress and ε is true strain. By this way W_C can be determined as the area under the true stress-true strain curve. (*Fig.20.*)

Gillemot stated that fracture occures in a material, when ASPEF will be absorbed independent of the kind of loading and of the geometry.

The basic concept of ASPEF is correct, because a deformation energy is always absorbed in a *given volume* of the specimen and not on the surface, as most energy concepts suggest. Therefore, ASPEF could be used to solve many different qualifying problems [10,11]. But ASPEF has also its limits. The effect of a complex stress state and the stress distribution within the cross section of a necked part of the specimen are not taken into account, although in some cases these factors have also a significant effect and should be regarded.

Radon and Czoboly recommended the use of ASPEF for calculating G_{Ic} or J_{Ic} [12]. The basic principle of this suggestion is the assumption that in the plastic zone ahead of a notch or crack W_c specific energy is absorbed before the crack starts to propagate. W_c is measured in [J/mm^3], which is physically correct. To get G_{Ic} or J_{Ic} (energy related to the surface) the multiplication of W_c with the length of the plastic zone L_c is necessary.

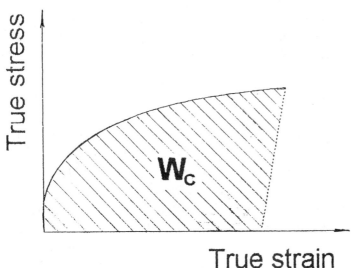

Figure 20. Determination of ASPEF as the area under the true stress/true strain curve.

$$G_{Ic} \text{ (or } J_{Ic}) = L_c . W_c \qquad (6.9)$$

where L_c is the width of the plastic zone at the crack tip represented by the length of an imagined, small specimen (*Fig. 21.*). A similar method was ellaborated by Havas with the help of COD [13]. By this methods J_{Ic} could be measured in many special cases, when other conventional procedures were not adequate[14].

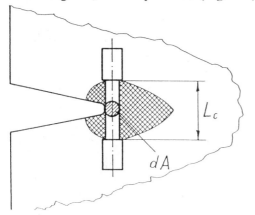

Figure 21. Imagined, small specimen at the tip of a crack or notch.

6.3. Impact Bending Test

Impact bending is also a very important and frequently used test in engineering practice, however, it has an inferior role for materials at high temperature. A notched specimen is broken with a blow by a pendulum and the energy needed to fracture is measured. The test has many advantages, as e.g. the simple test piece, the easy testing procedure and the quickly gained results. A great benefit is its sensitivity to the small changes in the structure due to varying technological parameters. But the test values, although undergoing a change in time, have not a correct physical meaning.

The basic idea that *energy* should be used for characterizing fracture properties is perfect, but the way how this energy is calculated is wrong. The first problem is that the measured quantity contains the total deformation work: both the crack initiating and propagating energies. These are functions among others of the shape and dimensions of the test piece. Particularly the notch depth and the notch radius have a decisive effect on the absorbed energy, therefore, the measured value is *not a material parameter* only a number characteristic on the *specimen.*

Another reason that this test value cannot be considered as a physically well defined parameter is the fact that the greatest part of the energy is absorbed in a volume, which is not specified in the test. The result of the test is expressed in J or formerly in J/cm². Neither of them can be regarded as real specific values.

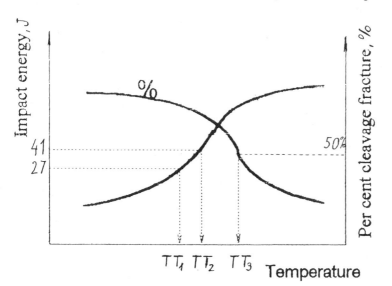

Nowadays the tests are performed as a function of temperature and the materials are classified by the *transition temperature,* where the behaviour of the test pieces changes from brittle to tough (*Fig.22.*). Since the transition is progressive, the criteria, where

Figure 22. Determination of transition temperature based on the usual criteria of impact bending test.

the transition temperature should be marked out is rather arbitrary. It can be done according to a given amount of the absorbed energy (27J; 41J;...), it can be based on the appearance of the fracture surface (50% crystalline-, 50% fibrous) or on the deformation of the specimen(1% lateral expansion at the rear edge of the specimen). Of course, all the possible criteria provide different TT values.

6.4. Fatigue Test

The detrimental effect of cyclic loading on the structures is well known to engineers for more than 150 years. Bookshelves can be filled with research reports and scientific papers, but an absolutely reliable designing method is still missing.

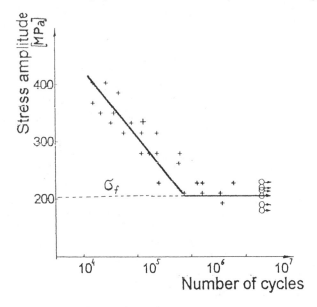

Figure 23. Typical fatigue curve of a low alloyed steel. σ_f - endurance limit.

The first testing method was recommended by Wöhler, who suggested to examine a set of specimens loading each test piece with a different, constant stress amplitude until the complete fracture. Typical results are shown in *Fig.23*. As illustrated also in the figure, there are so called "run-out" specimens too, which do not fracture within a given range of cycles. (In the case of structural steels this range was found to be 2÷3.10⁶.) According to observations no fracture could be expected afterwards. This provided the possibility to determine a so-called *fatigue limit*, as it can be seen in the figure.

This fatigue limit or endurance limit was used as a material parameter for many years. It was assumed that this value can be applied in dimensioning processes, the design will be safe and fatigue fracture will not take place, if the following assumption is fulfilled

$$\sigma_{equ} \le \sigma_f \qquad \text{instead of} \qquad \sigma_{equ} \le R_e \qquad (6.10)$$

where σ_{equ} is the equivalent stress acting on the component, σ_f is the fatigue limit and R_e is the yield stress. The practice has shown that this idea is incorrect and although design processses have been improved in the last decades, 80% of the technical failures are still associated in some sense with fatigue[15]. The reason of this lies in the poor physical content of the material characteristic.

Fatigue of materials is a rather complicated, complex process and a number of influencing factors have to be considered. To predict at least approximately their effects, the physical process has to be examined more closely.

Considering the process at first on an atomic and microscopic level, it is found that it starts with dislocation glide, although the stress is below the yield stress. The motion of the dislocations is restricted only to a few, favourably oriented grains (*Fig.24.*), namely to those, in which one of the potential slip systems is approximately parallel with the highest shear stress (45° inclined to the direction of tensile stress). Within these grains narrow slip bands are created consisting of a set of active slip planes close to each

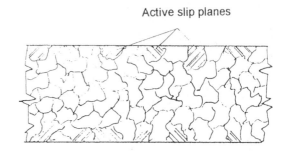

Active slip planes

Figure 24. Active slip lines in the favourably oriented grains of a crystalline material after alternating load less than the yield stress have been applied.

other. The distance between the active slip planes is about 100 a, where a is the lattice parameter. The density of slip bands within a grain is a function of the stress amplitude. The consequence of this strongly localised alternating gliding on the neighbouring, active slip planes is that so-called extrusions and intrusions (*Fig. 25.*) develop, which are the initiating sites of small, submicrocracks. The height of these ex- and intrusions from peak to valley has the magnitude of about 1000 a. Since the cracks are located within the slip bands they have the same orientation as the active slip system and their lengths are smaller than the diameter of the grains. This cracking process is called *Stage I cracking.*

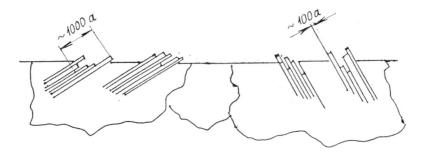

Figure 25. Extrusions and intrusions on a crystal due to cyclic loading. Submicroscopic cracks and voids nucleate at these points.

The extrusions and intrusions can most easily develop on the free surfaces. Therefore, fatigue fractures generally start on the outer surface or at the vicinity of an intrusion or other defect, which act as an inner surface. Beside the cracks, voids nucleate within the slip bands and by joining and combining with the submicroscopical cracks they grow to micro- and macro cracks. The direction of the cracks turn very soon normal to the principal tensile stress (*Stage II*) and after a longer or shorter period the process leads to final fracture [16]. The transition from a microcrack to a macrocrack can be defined by the help of fracture mechanics considerations, as shown below.

Every specimen examined in the frame of the Wöhler testundergoes the whole fatigue process from the first dislocation glide until the final fracture. That means that the effect of very different physical processes are mixed together in the test results. Therefore, the value of global information obtained from the tests is very limited. Furthermore, the tests are performed on smooth test pieces, without stress concentration, although it is well known that in practice fatigue fracture always starts at a notch or edge or similar sites, where the stresses are concentrated.

Because of the poor physical content of the endurance limit, this parameter is valid only for those unnotched specimens, which were applied in the test procedure and unlike the yield stress, this value cannot be transferred even to rods of different dimensions, still less to a real structure, with different shape, etc. The endurance limit is not even adequate to order the materials in a sequence. A material, which has a higher endurance limit than the other measured on smooth specimens, can be worse for a given purpose, because its notch sensitivity can be higher too.

Taking into account that almost all structures and machines are exposed to alternating loading and so a danger of fatigue failure exists, it is not astonishing that engineers and scientists searched for better solutions and in the last decades approached fatigue process in a different way. The Wöhler procedure and the endurance limit, as a material parameter lost its dominancy. Scientists realized that even if it is impossible to observe the whole fatigue process in details in most practical cases, at least the two basic phases should be divided: macro crack nucleation and crack propagation.

In handling these two processes two different theories were developed. The first one considers crack initiating period as a process of repeated plastic deformations, which can be modelled by an appropriate test (low cycle fatigue test) [17]. The second one regards the nucleation period also as a crack propagation process, however, the general laws relevant to the macrocracks are not valid here [18].

According to the first theory crack nucleation can be determined by *low cycle fatigue* data, because it is justified to assume that the crack will start at some stress concentrator, where the stress surpasses the yield point and so plastic deformation will occure in every load cycle. The local strain at such a concentrator can be determined according to the method suggested by Neuber [19]. On the one hand the Neuber hyperbol should be constructed

$$(\sigma_{max} \cdot \varepsilon_{max}) = (K_t \cdot \sigma)^2 / E \qquad (6.11)$$

where σ_{max} and ε_{max} are variables for the peak values at the notch tip and K_t is the notch factor and on the other hand the cyclic stress-strain curve should bee plotted (*Fig.26.*). According to Neuber's theory the real σ_{max} and ε_{max} values are indicated by the crossing point of the two curves.

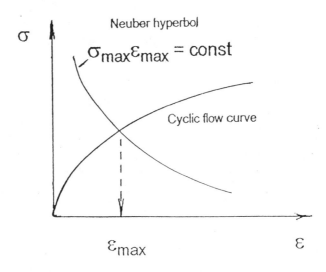

Figure 26. Determination of ε_{max} with the help of Neuber hyperbole and the cyclic flow curve.

Getting the strain value at the notch tip, the effect of this cycling strain is simulated by a test, where the specimens will be plastically deformed in every cycle. For convenience, the tests are usually performed by strain control. A typical stress/strain loop for a low cycle fatigue test is shown in *Fig 27*. The shape of the loops changes during the test, at first because of softening or hardening processes and close to the end of the life, because of the effect of a developing macrocrack.

With a set of specimens a diagram similar to the Wöhler curve can be constructed where the strain amplitude is plotted against the number of load cycles in a double logaritmic scale. Such a diagram - the Manson-Coffin curve - can be

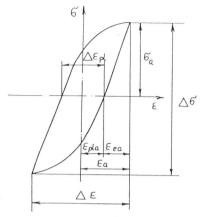

Figure 27. Stress/strain hysteresis loop presenting the characteristic quantities.

seen on *Fig. 28*. With the help of this diagram the first part of the fatigue life: the number of cycles necessary to initiate a macro crack, N_i can be estimated. For this purpose the peak strain calculated by the Neuber's theory, ε_{max} has to be introduced into the diagram, as shown in Fig.28.

The results are more realistic, if the fracture criteria in the Manson-Coffin diagram is not the total fracture of the specimens, but e.g. the appearance of a 2 mm crack. A usual method in low cycle fatigue test is to continue the tests until the tensile load decreases with a given percent due to a crack. This fits much better to the real cases, because a crack about this size can be regarded as a macrocrack. Nevertheless, low cycle specimens are also smooth, unnotched test pieces, where the stress distribution is uniform through the cross section [20]. It was observed [21], on the contrary, that the stress gradient has an important influence on the behaviour of the specimen, which observation is however, neclected.

After the determination of the first part of lifetime, the second part of the fatigue life has to be calculated as a *crack propagation* process. Although, the gliding mechanism of dislocations is still continued and some other small cracks may nucleate in other parts of the specimen, the deciding process with a view to the fatigue life is the propagation of the main crack. This is in most cases a

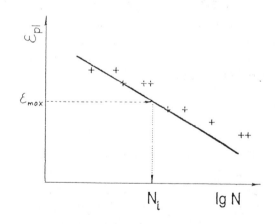

Figure 28. Manson-Coffin diagram obtained by low-cycle tests. The first part of fatigue life (crack initiation) can be determined, if the peak strain is known.

transgranular crack and its global direction is normal to the principal stress [22]. The rate of propagation is in the beginning slow, but it accelerates with the number of cycles. (*Fig. 29.*)

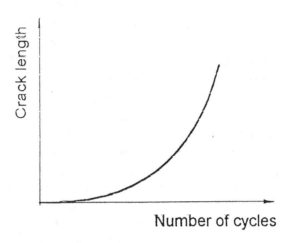

Figure 29. Schematic illustration of fatigue crack propagation vs. number of load cycles.

A large number of different formulae have been recommended to describe crack propagation. The greatest probleme here was to find a method, which provides the possibility to transfer the test results to the real structures. As mentioned earlier, fracture mechanics is today the only possible way to handel cracks in such a sense that the results can be generalized. More of the suggested formulae used fracture mechanics terms, but the most simple empirical equation was given by Paris and Erdogan [23].

$$dc/dN = C. \Delta K^m \tag{23}$$

where dc/dN is the rate of crack growth per cycle, ΔK is the range of stress intensity factor depending on the crack size, c and on the range of stresses, $\Delta\sigma$. C and m are material characteristics. During a test ΔK is steadily increasing, assuming that a macrocrack has been developed, which is steadily growing. Using a double logaritmic scale, Paris-Erdogan law provides a straight line, as seen in *Fig. 30.* However, the measurements validated the formula only partly, as seen on the figure, because the linear relationship is not true either at low, or at high ΔK values.

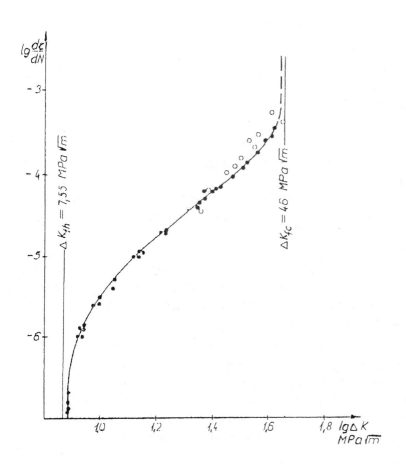

Figure 30. Fatigue crack propagation as a function of stress intensity range.

According to many test results the crack does not propagate below a characteristic stress intensity range, nominated as the *threshold stress intensity range*, ΔK_{th}. On the contrary, it speeds up near to the final fracture, which occures at a critical K_{fc}. Although, other formulae recommended in the literature reflect more exactly the process of crack propagation, these are generally not used, because they are far too complicated.

The lifetime of crack propagation period can be calculated by integrating Equ.23 according to dc and dN, resp. The limits of integration are c_o, the length of the initial macrocrack and c_{crit}, the critical crack length, calculated from K_{fc}.

An exact determination of K_{fc} is however problematical. Theoretically fracture would occure, if K_{Ic} is reached. But K_{Ic} belongs to a "virgin" material, which did not experienced plastic deformation before. In fracture mechanics tests the size of plastic zones are restricted to a maximum value, but which can be easily surpassed in the case of fatigue. Mind, that the crack is growing slowly until the final, critical size. Cyclic plastic deformation occurs before the crack tip during each cycle. This procedure may alter the original properties of the material and so it may influence the value of the critical stress intensity factor.

There are even other problems with the calculation of crack propagation period. The number of cycles needed to increase the crack from c_o to c_{crit} depends much more on c_o than on the size of the critical crack. But as shown earlier, just the size of the initial crack is uncertain, because it is rather arbitrary what is called as a macrocrack.

Furthermore, there are problems with the Paris-Erdogan equation itself. Although, the *C and m* parameters can be determined with a reasonable accuracy and this strengthens the reliablity of the calculations, but the parameters have no physical meaning, since the Paris-Erdogan equation is absolutely empirical. The dimensions of the left and right sides match only in that case, if C has a complex dimension: $[mm^{5/2m}/N^m.cycle]$. The integration based on dc makes it even more complicated. Further more the given formula describes the process only in the intermediate range, as it was shown. A further disadvantage of the Paris-Erdogan law is that the mean stress, an important influencing factor, is not included. However, the method is comparatively simple and it is generally used, because the measured data fit well on the curves.

Finally a few words should be said about the other scientific school, which regards the whole fatigue process as a continuous crack propagation [25]. Of course, the way of approach concerning the macro crack is the same, as described before, but the nucleation of the crack is explained otherwise. According to this oppinion there are small defects in any material and due to the alternating load they start to propagate. Such small defect is probably only an array of dislocations.

Now the question may be asked, how can these small cracks grow, when the stress intensity range is below the threshold. The answer is that the small and short cracks cannot be treated by linear elastic fracture mechanics and its parameters, because these are macroscopic concepts and are based on elasticity theory. The small cracks, whereas have a definite crystallographic orientation, they are located in a slip band, so their stress field around the crack tip cannot be characterized by the stress intensity factor.

Experimental observations have proved that the propagation of these submicro cracks is really absolute different from the larger ones. To the contrary to the general law that with increasing crack length - which means increasing stress intensity - the rate of propagation is also growing, the rate of expansion of the small cracks *decreases* with crack growth. As shown by dotted lines in *Fig. 31.* the velocity of crack propagation reaches a minimum, which can also be zero. In this case the crack stops. '"Non-propagating cracks" are well known for many years. Otherwise, the propagation accelerates and finally reaches the value prescribed by the macroscopic law.

This behavior of cracks provide also the possibility to define a "macrocrack". A defect can be regarded as a macrocrack, if the propagation follows the rule, given by the dc/dN - ΔK curve.

Two more problems should be shortly mentioned here without a detailed discussions: Succeeding from the probabilistic nature of the fatigue process, the results scatter very much. Therefore, a statistical processing is advantageous, but this needs a relatively great series of specimens tested, which are not always available.

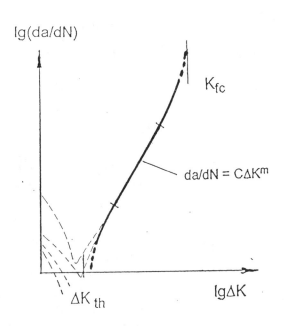

Another frequent problem is the elaboration of random loading [26]. The damaging of the material by an individual load cycle depends not only on the cycle itself, but also on the previous load history. Various cumulative damage laws have been recommended in the literature, but none

Figure 31. Propagation of "short cracks", which do not obey the law for macrocracks.

of them proved to be universal. Mostly the simplest law, Miner's rule will be applied, although it is well known that the results can be very misleading.

6.5 Creep Tests

First of all the standard tests, which are performed to determine the load-carrying capacity of materials at elevated temperature should be discussed. The word "elevated" can mean very different temperatures depending on the material; generally it is above 0,5 T_m, where T_m is the melting point in Kelvin degrees. At lower temperatures the maximum permissible load is determined by the yield stress. Creep will occure only at a higher stress. However, with increasing temperature creep limit decreases faster, than the yield stress(*Fig. 32.*).

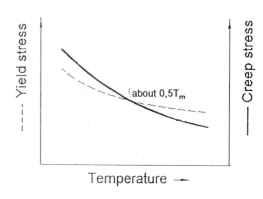

The basic methods use a series of specimens loaded by constant loads and measuring the time to rupture (*stress rupture test*), or the deformation versus time (*creep test*) or both (*creep-rupture test*). In the background of these tests stands the necessity to build structures, which

Figure 32. Variation of warm yield stress and creep limit vs. temperature.

should be in service for many years (some of them for more than 30 years) without fracture or any other disturbances, like an undesirable permanent deformation. In elements of power stations e.g. 1% permanent strain is permitted after 100,000 hours service.

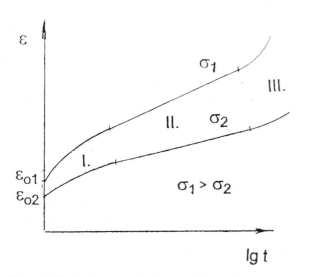

Figure 33. Strain as a function of time and stress. Stage I - transient creep, Stage II - steady state creep, Stage III - tertiary creep.

Typical behaviour of metal specimens exposed to tensile stresses at high temperature are illustrated in *Fig. 33.* The rate of strain and the time till fracture depends on the applied stress and on the temperature. But in general case the character of the curves is the same: three parts can be distinguished, denoted by Stage I, II and III creep. The Stage III creep may be missing, if Stage II creep will be zero. In other case, if Stage II creep has a greater than zero value, Stage III creep leads to fracture. Therefore, the zero creep rate seems to be very promising regarding real structures, where fracture should be avoided. However, the determination of the limit stress is difficult and the application of such a stress limit is not economical.

As described in Chapter 4. more mechanisms can be active at elevated temperature and result plastic deformation. The governing mechanism depend on the stress level and on the temperature, but generally all mechanisms contribute to the total strain. The ratio of contribution changes also as a function of time even at constant stress and temperature. This is the explanation of the different character of the creep curve as a function of time.

At loading the specimen will be deformed immediately, the strain (elastic and possibly plastic) being proportional with the stress. Afterwards the creep mechanisms become active and will deform the specimen further.

Mathematical description of the process is difficult, because of the interacting mechanisms. The strain is usually given by the formula:

$$\varepsilon = \varepsilon_0 + \Sigma a_i t^{-m_i} + \Sigma b_j t^{-n_j} \tag{24}$$

where ε_0 is the instant strain, t is time and a_i, b_j, m_i and n_j are stress and temperature dependent material constants[4]. More often the strain is described by empirical formulae for the individual sections of the curve.

In Stage I creep (named also transient creep) strain rate decreases continuously. The most important mechanism in this section is thermal activated glide of dislocations. The number of dislocations increases and in many materials they form a subgrain structure. As a result of this motion, the number of mobile dislocations decrease, or with other words: work hardening occures. The strain versus time can be described by one of the equations:

$$\varepsilon = \varepsilon_0 + \alpha.\log(1+\gamma t) \tag{25}$$

$$\varepsilon = \varepsilon_0.\beta.t^{1/3} \tag{26}$$

where α, β and γ are material constants. According to observations, Equ. 25 is valid for lower temperatures, while Equ.26 fits more the experimental results at higher temperatures.

The most important part of creep is Stage II., where the creep rate has a minimum value. The strain rate is constant for a longer period, which shows that an equilibrium state has been developed. Gliding of dislocations continues, but the work hardening process is balanced by a "softening" effect. This is due to the nucleation of a great number of vacancies, according to the temperature. The interaction of the vacancies and dislocations result non-conservative motion of dislocations, which keeps the number of mobile dislocations constant.

The phenomenally discription of the steady-state creep is simple:

$$\varepsilon = \varepsilon_0 + \varepsilon\, t \tag{27}$$

where ε can be described by various formulae [4]. Some of these are:

$$\varepsilon = A\sigma^{m}\exp(-Q/RT) \tag{28}$$

$$\varepsilon = B\sigma^{n} \tag{29}$$

$$\varepsilon = C.\sinh(\sigma/\sigma_{0}) \tag{30}$$

$$\varepsilon = D.\exp[-Q+f(\sigma)/RT] \tag{31}$$

In Equs.28-31. *A, B, C, D, n, m* and σ_{0} are material constants dependent on temperature and *Q* is the activation energy of creep. It has been shown that this quantity is nearly equal to the activation energy of self diffusion. This is an indirect evidence that self diffusion (the presence of vacancies) control stready state creep [2].

Although the creep processes are balanced during Stage II creep, some permanent damage occures within the structure, which leads finaly to the acceleration of deformation. Earlier it was thought that the reason of this increased creep rate is mainly the increase of true stress due to the constant load and to the decreasing cross section. Further experiments have demonstrated that the tertiary creep follows the steady state creep inevitably, if the strain rate is greater than zero. The real reason of Stage III creep lies in serious damage processes in the structure. As discussed in Chapter 4. plastic deformation, mainly the stress directed diffusion and the crack boundary slip have a consequence of void and microcrack nucleation and this leads finally to rupture. Other alterations in the microstructure, as onset of recrystallisation, or rearrangement of second phase particles may influence strongly tertiary creep.

It is worth while to mention, that the total strain till fracture may decrease drastically with reduced stresses. The role of voids and cracks at the grain boundaries may dominate in the case of low stresses and high temperatures, when the strain rate is low too and the fracture will transform from transgranular to intergranular type. In this case brittle fracture may occure in a macroscopic sense.

However, in general creep process is connected with plastic deformation and so, the physical content of the creep parameters is correct. They describe the behaviour (the elongation) of the specimen under the influence of a given stress, at a given temperature. The creep process is really related to these factors. Every part of the specimen, which is exposed to these conditions will take part in the process. The effect of multiaxial stress state can be calculated by using the general yield theories. This is true as long as the cracks do not disturb the stress distribution in the macroscopically quasi homogenous material. If cracks are present, fracture mechanics methods should be applied.

It is more problematic to relate the measured strain to the different creep mechanisms, since the total strain is always a product of more than one mechanism. The contribution of the individual mechanisms to the total strain is a function of the stress and the temperature. The deformation mechanism maps (*Fig.34*) can help in orientation. On the axes of the map are the normalized stress, σ/G and the homologous temperature, T/T_m. The map shows those stress and temperature combinations, where one or other deformation mechanism is dominant. The lines represent those points, where the contribution of two, or three mechanisms are about the same.

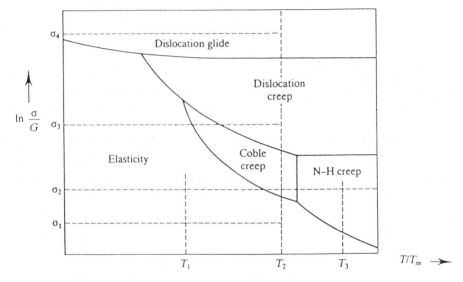

Figure 34. Schematic illustration of deformation mechanism map [2].

The greatest problem in creep and creep-rupture measurements is the *time*. To complete tests mainly at elevated temperatures for 100,000 hours or more is very expensive and troublesome. In addition, the results are generally needed in a much shorter period. Therefore, different methods are recommended in the literature to save time, that is to accelerate the tests. One of the most frequently applied method is the use of the Larson - Miller parameter. The authors considered creep as a process, controlled by diffusion. According to Equ.28. a given amount of strain can be produced in time *t*.

$$\varepsilon = \varepsilon t = A\sigma^m exp(-Q/RT).t \tag{32}$$

Taking the logarithm of both sides

$$\ln\varepsilon = \ln A + m.\ln\sigma - (Q/RT) + \ln t \tag{33}$$

Considering that a given strain should be produced, the right side of the equation is also constant. Keeping the stress also invariant, all the constant quantities can be reduced to one constant, C. Rearranging the equation, we obtain:

$$Q/R = T(C + \ln t) \tag{34}$$

Since Q can be regarded as a constant value, if the conditions are about the same (self diffusion controls creep, there are no precipitation or solution processes, etc.) Equ.34 means that temperature and time are in some sense equivalent to produce a given amount of strain at a given stress level. Taking logarithm to the base 10 the so called Larson-Miller parameter is obtained.

$$T(20 + \log t) \tag{35}$$

where T is temperature in Kelvin, t is time in hours and C is between 17 and 23 for metals. By using this parameter, time can be spared by raising the temperature. Because of possible metallurgical changes (e.g.: dissolution of precipitates, recystallization or fase transformation), this method can be used only with known groups of materials and in a limited range of temperature. In most cases temperature increase is about 50° C. If void- and microcrack nucleation takes place within the lifetime of the test and a considerable part of the deformation originates from this processes, the extrapolation is not justified. Anyhow, the shorter the time of test, t_f, compared to the sevice life, t_s, the more uncertain are the results, as indicated on *Fig. 35*.

Other methods are also recommended in the literature, but because of the complexity of the deformation mechanisms there is no generally accepted process.

6.6. Combined Test Methods

Because the loading conditions are in practice much more complicated than the simple loading models (test methods) discussed in Chapter 6., complex tests are often necessary. In such cases the combined effect of temperature, corrosive environment, radiation, constant and alternating loading, etc. are examined. The results are generally evaluated in terms of empirical equations and empirical material constants.

It is almost impossible to overview all the possible combinations of interior (different phases, grain size, inclusions, texture...) and exterior (temperature, multiaxial stress state, mean stress, strain rate, environment...) factors, which have an effect on the individual processes discussed before and which may influence the lifetime of a part. Many of the factors are analysed in Ref. [27]. Here the discussion will be restricted to the effect of temperature and alternating stress as one of the usual combinations in practice. In studying the combined effect two ways are possible. The analysis may be based on the fatigue process studying the influence of elevated temperature. Or one may examine the creep processes as a basis and investigate the effect of alternating loading. It depends on the actual case, on the actual parameters, which of the two methods is more justified.

The measured quantities, as well as the criteria of failure are also different in the two cases. In the fatigue experiments the number of load cycles, the rate of crack propagation or even total fracture is observed. Otherwise the strain until a given time, the strain rate, or the lifetime should be measured. If crack propagation process is included, the concepts of fracture mechanics should be used. Typical experiments are the examination of crack propagation rate, expressing the results by the Paris-Erdogan equation and investigating the influence of the special test conditions on the C and m constants. However, at high temperature general plastic flow may occure, which should be excluded basically from fracture mechanical analysis. In such cases the crack propagation law is somewhat altered and instead of ΔK similar parameters, as C^* or C_t are applied [28-30].

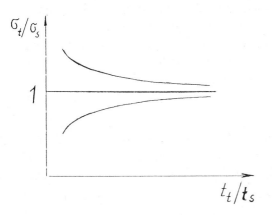

Figure 35. Uncertainty of the test results increases, if the time of tests decreases compared to time of service.

If cyclic loading is applied at high temperature, more exactly above $0.5\ T_m$, the physical process of fatigue will be combined with the processes of creep. As discussed earlier, at this temperature non-conservative motion of dislocations, stress directed diffusion, boundary diffusion and void nucleation at the grain boundaries are possible. This procedures have a strong influence on fatigue mechanism. The climbing mechanism of dislocations will result that the originally planar arrangement of dislocations, which characterises the first period of fatigue, will turn to a so-called *wavy* distribution. It was observed that wavy arrangement of dislocations leads much earlier and easier to a Stage II crack than usual.

The combination of damageing mechanisms of fatigue and creep has a drastic effect on the lifetime. It has been discussed earlier that the first stage of fatigue life is crack initiation. This mechanism is enhanced very much by void nucleation on the grain boundaries. The existance of voids promote also crack propagation, which will run also along the boundaries. So, to the contrary of usual transgranular fatigue cracks at low temperature, the fracture may occure intergranular at high temperatures. Intergranular fracture develops under appropriate circumstances more easily, than transgranular, which means that the fatigue life can be drastically reduced.

Because all the creep mechanisms are dependent on time, the dominating factor in fatigue will be a function of time too. Of course, not the total time, but only those periods have a role, when a relatively high stress is acting. So, e.g. the significance of wave form of the load cycle will increase, or - which is the same - the hold time at the peak load will be important. The same concerns also to frequency. Low frequency means that untill a given number of cycles a long time passes in contrast with a high frequency. Therefore, creep mechanisms have time to contribute to the damage of materials. Data can be found, that the reduction of frequency from 600 to 2 cpm results a decrease of the fatigue life by a factor of 100 [16].

The interaction of creep and fatigue can be examined also differently: taking the process of creep as a primary one and investigating the effect of varying load. Such investigations have shown that the minimum strain rate for a cyclic loading is in the most favourable case equal to the creep rate under static load, assuming that the maximum stress in the load cycle is equal to the static stress in the creep test. However, in many cases the minimum creep rate under cyclic loading is higher [31].

The observations are similar for the lifetime for creep. The life is reduced, if the stresses are not constant, but vary cyclically. The reduction depend on the material and on the loading parameters. In the most favourable case, the reduction is negligable, but the resultant lifetime is never longer than in pure creep.

7. REFERENCES

1. Czoboly, E.(1995) *Overview on monotonic, creep and cyclic stress strain behaviour at high values of strain. In:* Mechanical behaviour of materials at high temperature. Ed.:C. Moura Branco, R.Ritchie and V.Sklenicka. NATO ASI Series. Kluwer Academic Publisher. Dordrecht /Boston/London. 3-22.

2. Courtney, Th. H. (1990) *Mechanical behaviour of materials.* McGrow-Hill Publishing Company. New York, USA

3. Schatt, W. (1972) *Einführung in die Werkstoffwissenschaft.* Deutscher Verlag für Grundstoffindustrie. Leipzig, Germany

4. Finnie, I. and Heller, W.R. (1959) *Creep of Engineering Materials.* McGraw-Hill Company, Inc. New York/Toronto/London.

5. Irwin, G.R. (1958) *Fracture mechanics.* Contr. to the First Symp. on Naval Struct. Mechanics. Stanford University, Stanford, USA.

6. Griffith, A.A. (1920) *The phenomenon of fracture and flow in solids.* Phil. Trans.Roy. Soc. London, A-221, 163-179.

7. Berkovic, M., Sedmak, A. and Jaric, J. (1990) *C* integral - theoretical basis and numerical analysis.* Proc. 5th Int. Fracture Mechanics Summer School. EMAS, Warley, U.K. 71-88

8. Ginsztler, J. (1989) *Retarding the crack initiation process during low cycle thermal shock fatigue.* In: Low Cycle Fatigue and Elasto-Plastic Behaviour of Materials. Ed.: K.T.Rie. Elsevier Applied Science, London 643-648.

9. Gillemot, L. (1963) *Die Beurteilung der Werkstoffe auf Grund der Brucharbeit.* Freiberger Forschungshefte September, 5-13.

10. Gillemot, L. (1963) *Beurteilung der Schweissbarkeit an Hand der Brucharbeit.* Schweisstechnik **13**. H.7. 305-312.

11. Czoboly, E., Havas, I. and Gillemot, F. (1982) *The absorbed specific energy till fracture as a measure of the toughness of metals,* Proc. Int. Symp.on Absorbed Specific Energy and/or Strain Energy Density Criterion Akadémiai Kiadó, Budapest, Hungary **3,** 330-336.

12. Radon, J.C., Czoboly, E. (1972) *Material toughness versus specific fracture work.* Proc. Int. Conf. on Mechanical Behaviour of Materials, Kyoto, Japan, 543-557.

13. Havas, I., Schulze, H.D., Hagedorn, K.E. and Kochendörfer, A. (1974) *Der Zusammanhang zwischen der spezifischen Brucharbeit und der Bruchzähigkeit.* Materialprüfung **16**. Nr.11.349-353.

14. Czoboly, E. et al. (1989) *Fracture mechanics concept to increase the reliability of high-pressure gas cylinder.* Proc. ICF7. Advances in Fracture Research. Pergamon Press, New York, USA, 3555-3562.

15. Dauskardt, R.H. and Ritchie, R.O. (1993) *Fatigue of advanced materials. Part I.* Advanced Materials & Processes, July, 26-31.

16. Gell, M. and Leverant, R. (1973) *Mechanisms of high-temperature fatigue.* In: Fatigue at Elevated Temperatures. Ed.: A. E. Carden, A. J. McEvily and C. H. Wells. ASTM Publication 520. Philadelphia, PA, USA. 37-66.

17. Benham, P.P. (1958) *Fatigue of metals caused by a relatively few cycles of high load or strain amplitude.*Metallurgical Reviews **3**.No.11. 203-234.

18. Miller, K.J. and de los Rios, E.R. (1986) *The Behaviour of Short Fatigue Cracks.* European Group on Fracture Publication 1. MEP Institution of Mechanical Engineers, London, UK.

19. Neuber, H. (1961) *Theory of stress concentration for shear-strained prismatical bodies with arbitrary nonlinear stress-strain law.* Trans ASME, Series E., **83**

20. Czoboly, E., Havas, I. and Ginsztler, J. (1984) *Relation between low cycle fatigue data and the absorbed specific energy.* Proc. 5th EGF. Lisbon, Portugal, EMAS, Warley, U.K.481-494.

21. Czoboly, E. and Sandor, B.I. (1974) *Fatigue behaviour of notched steel specimens.* EES Report No 39. University of Wisconsin, USA 1-236.

22. Broek, D. (1978) *Elementary Engineering Fracture Mechanics.* Sijthoff and Noordhoff, Alphen aan den Rijn, The Netherlands.

23. Paris, P.C. and Erdogan, F. (1963) *A critical analysis of crack propagation laws.* J. Basic Eng. Trans ASME Series D. **85**. 528-534.

24. Tóth, L. (1994) *Reliability assessment of cracked structural elements under cyclic loading.* In: Handbook of Fatigue Crack, Ed.: A.Carpinteri. Elsevier, Amsterdam1643-1683.

25. Miller, K.J.(1991) *Metal fatigue - past, current and future.* Proc. of Institution of Mechanical Engineers, **205** 1-14.

26. Radon, J.C. and Czoboly, E. (1988) *Problems of fatigue crack growth measurement under random load.* Periodica Polytechnica **32**. No.2. 107-117.

27. Radon, J.C. (1990) *Fatigue crack propagation at elevated temperatures.* Proc. 5th Int. Fracture Mechanics Summer School. EMAS, Warley, U.K. 117-134.

28. Webster, G.A. (1996) *Creep and creep fatigue crack growth of high strength steels* In: Mechanical behaviour of materials at high temperature. Ed.:C. Moura Branco, R.Ritchie and V.Sklenicka. NATO ASI Series. Kluwer Academic Publisher. Dordrecht /Boston/London. 169-193

29. Saxena, A. et al. (1990) *Application of C_t in characterizing elevated temperature crack growth during hold time.*In: Elevated Temperature Crack Growth .Eds.: S.Mall and T.Nicholas.ASME Book No G00530.

30. Hollstein, T. and Kienzler, R. (1990) *Creep crack growth experiments and their numerical simulation.* Proc. 5th Int. Fracture Mechanics Summer School. EMAS, Warley, U.K. 175-188.

31. Lukas, P., Kunz, L. and Sklenicka, V. (1996) *Interaction of high temperature creep with high cycle fatigue.* In: Mechanical behaviour of materials at high temperature. Ed.:C. Moura Branco, R.Ritchie and V.Sklenicka. NATO ASI Series. Kluwer Academic Publisher. Dordrecht /Boston/London. 155-167.

HIGH TEMPERATURE DEFECT ASSESSMENT PROCEDURES

R.A. Ainsworth
Nuclear Electric, Barnwood, UK

ABSTRACT

In this chapter, a basic high temperature defect assessment procedure for components subjected to essentially steady load is described. This is based on the Nuclear Electric R5 procedure, the British Standards Document PD6539 and other procedures in the literature. First, the basic phenomena influencing high temperature defect growth are discussed and the basic calculations needed and the input materials data required to assess these phenomena are presented. Then this information is combined into an overall defect assessment procedure which is discussed. This procedure is finally illustrated by a number of worked examples. The presentation here is necessarily brief and for more detail on the background phenomena, assessment procedures and worked examples the interested reader is referred to the book by Webster and Ainsworth [1].

1.   OVERALL BEHAVIOUR

The behaviour of a crack characterised by initial dimensions a_o, l_o (the depth and semi-length of a semi-elliptical surface defect, say) is shown schematically in Figure 1. One factor which limits the allowable size of this defect, either initially or after some creep crack growth, is short-term fracture. Methods for assessing this mode of failure are discussed first in Section 2, including estimation of the post-yield fracture mechanics parameter J.

Assuming the initial crack does not fail by short-term mechanisms, creep will affect subsequent response as shown in Figure 1. Initially creep straining may lead to blunting of the crack tip without significant crack extension. The duration of this initiation period, t_i, is discussed in Section 3. This includes methods for estimating the steady state creep parameter C* which is the creep equivalent of J. The transient period prior to a steady state creep stress distribution being established is also discussed. It is shown that the initiation time may be correlated directly with C* or may be related to the incubation crack opening displacement, δ_i, which governs when crack growth starts, Figure 1.

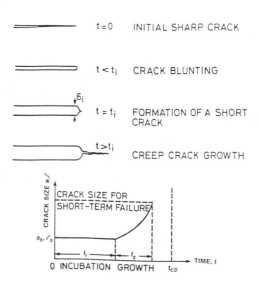

Figure 1 Schematic Defect Behaviour at High Temperature

The duration of the creep crack growth period, t_g, is discussed in Section 4 in terms of the associated creep crack growth rate, $\dot{a}$. The total failure time for initiation and growth is then simply $t_i + t_g$. However, this estimate of failure time may not correspond to the limiting failure mechanism. Failure may, instead, be governed by continuum damage accumulation in the ligament ahead of the crack. Estimates of the corresponding failure time, t_{CD}, are given in Section 5.

In Section 6, the various calculations in Sections 2-5 are combined into an overall defect assessment procedure. This is a simplified version of procedures in Webster and Ainsworth [1], R5 [2] and PD6539 [3]. Use of either parts of this procedure or the overall approach are then illustrated in Section 7 by means of worked examples.

2.   ASSESSMENT OF SHORT-TERM FRACTURE

2.1   Numerical Estimates of J for Power-Law Materials

For a material which deforms according to power law plasticity with strain, ϵ, related to stress, σ, by

$$\epsilon = \alpha \epsilon_Y \, (\sigma / \sigma_Y)^N \qquad (1)$$

where $\alpha \, \epsilon_Y$, σ_Y and N are constants, then the crack tip characterising parameter J for a defective component subject to a load P is

$$J = \alpha \, \sigma_Y \, \epsilon_Y \, c \, h_1 \, (P/P_Y)^{N+1} \qquad (2)$$

where c is a characteristic dimension, P_Y is a convenient normalising load proportional to σ_Y and h_1 is a non-dimensional function of geometry, crack size, mode of loading and N. The form of equation (2) follows from dimensional arguments since for a material deforming according to equation (1), the stress distribution is directly proportional to load, P, and the strain distribution is proportional to P^N.

Numerical solutions for J for a number of geometries, crack sizes and values of N have been obtained by Kumar et al [4]. These have then been tabulated in terms of the non-dimensional quantity h_1 in equation (2), thus enabling J to be evaluated provided the material can be adequately described by equation (1). For the centre-cracked plate geometry shown in Figure 2, these solutions are shown graphically in Figure 3.

It is apparent from Figure 3 that the normalised solutions are strongly dependent on both stress index N and on crack size. This makes the results difficult to use if there is uncertainty about the material description or interpolation for crack size is required.

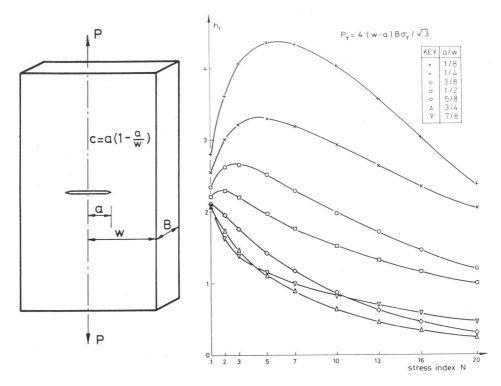

Figure 2 Centre Cracked Plate   **Figure 3 Variation of h_1 with N**

To overcome these difficulties, re-normalisation of the solutions in [4] has been examined by Ainsworth [5]. Essentially, the value of J is independent of the way in which it is normalised, specifically it is independent of the choice of normalising load in equation (2). If a different normalising load, P_1 say, were chosen then J could be written

$$J = \alpha \, \sigma_Y \, \epsilon_Y \, c \, h_1^1 \, (P/P_1)^{N+1} \qquad (3)$$

with the normalised parameter h_1^1 related to h_1 by

$$h_1^1 = h_1 \, (P_1/P_Y)^{N+1} \qquad (4)$$

to ensure that equations (2) and (3) give the same result. Results for h_1^1 are shown in Figure 4 for various values of P_1 for a compact tension specimen in plane strain with a crack size equal to half the section width w. The value P_Y is that used in [4] and leads to a strong variation of h_1 with N. It is apparent from Figure 4 that there is a specific normalising load P_1^* which makes h_1^1 sensibly independent of N. Evaluation of P_1^* requires numerical solutions for J to be available for a range of N values. Of more interest in Figure 4 are the results when P_1 is chosen equal to the plastic collapse load of the cracked geometry, P_{Lc}. This leads to a modest variation of $h_1{}^1$ with N, a factor of about 2 for $1 \le N \le 20$ for the compact tension specimen, with the value at $N = 1$ providing a conservative estimate for higher values of N. This is a useful result as solutions for plastic collapse load are much more widely available than power-law solutions for J and this is used in Section 2.2 below to develop approximate reference stress estimates of J.

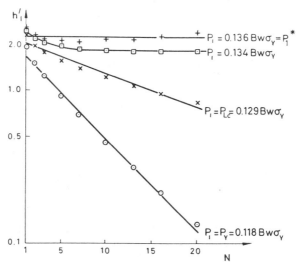

Figure 4 J Solution for Compact Tension Specimen using Different Normalisation

Renormalisation using the plastic collapse load is also useful for rationalising solutions at different crack lengths. For example, for a plane strain single edge cracked plate in tension, solutions for h_1 are given by Shih and Needleman [6] for $N = 16$ as 34.1, 1.90, 0.225 for crack depths $a/w = 0.125, 0.25, 0.375$ respectively. This strong dependence on a/w makes it difficult to interpolate to obtain solutions at intermediate values of a/w. However, if the plastic collapse load, P_{Lc},

$$P_{Lc} = 1.155 \, BW\sigma_Y \, [1 - a/w - 1.232 \, (a/w)^2 + (a/w)^3] \qquad (5)$$

for this geometry is used rather than the normalising load P_Y of [6],

$$P_Y = 1.455 \, Bw \, \sigma_Y [\{(1 - a/w)^2 + (a/w)^2\}^{1/2} - a/w] \qquad (6)$$

then a weak dependence on a/w results. This can be fitted [1] by a quadratic function in (a/w), leading to

$$h_1 = [11 - 56 \, (a/w) + 84 \, (a/w)^2] \, (P_Y/P_{Lc})^{17} \qquad (7)$$

which gives values of 33.7, 1.87, 0.222 for crack depths of a/w = 0.125, 0.25, 0.375 respectively, very close to the numerical results of [6] given above. Equation (7) is used in Section 7 to enable a range of crack sizes to be readily assessed.

2.2   Reference Stress Estimates for J

Equation (1) with N=1 and $\sigma_Y/\alpha\epsilon_Y =. E$, Young's modulus, corresponds to an elastic material with Poisson's ratio, $\nu = 0.5$. Therefore, for this case equation (3) must give a value for J equal to the elastic value G. Taking P_1 as the plastic collapse load P_{Lc} then the corresponding value of h_1^1 from equation (3) must satisfy

$$h_1^1 = EG/[c\ (P\sigma_Y/P_{Lc})^2]\tag{8}$$

$$=EG/[c\ \sigma_{ref}^2]\tag{9}$$

where

$$\sigma_{ref} = P\sigma_Y/P_{Lc}\tag{10}$$

is a reference stress. Recalling from Figure 4 that this value of h_1^1 at N=1 is an overestimate of values at higher N values, then a conservative value for J for N>1 follows from equations (3, 9, 10) as

$$J/G = E\ \epsilon_{ref}/\sigma_{ref}\tag{11}$$

where ϵ_{ref} is the strain at the reference stress from equation (1). The accuracy of equation (11) has been assessed in [7] for the range of solutions in [4] for which J solutions are available. Generally, equation (11) has been found to be conservative with a typical conservatism of about 5% on load; ie. the value of J obtained from equation (11) at a load P is typically equal to that from equation (2) at a load 1.05P. Equation (11) does not depend on any of the constants in equation (1) and can, therefore, also be used to estimate J for materials which do not obey power-law plasticity. This is a powerful result as it enables values of J to be simply estimated for any material, crack size and geometry provided an elastic solution and a limit load solution are available.

Equation (11) is conservative in the fully plastic region; it is also clearly accurate for elastic behaviour ($\epsilon_{ref} = \sigma_{ref}/E$, J = G). However, it needs to be modified in the small-scale yielding region ($\sigma_{ref} < \sigma_Y$) where J/G>1. These modifications generally correspond to a modest increase in J and a convenient correction to equation (11) is

$$\frac{J}{G} = \frac{E\ \epsilon_{ref}}{\sigma_{ref}} + \frac{\frac{1}{2}\ (\sigma_{ref}/\sigma_Y)^2}{(E\ \epsilon_{ref}/\sigma_{ref})}\tag{12}$$

This provides a correction at small loads which is phased out as the fully plastic term (the first term on the right hand side of equation (11)) becomes large. Equation (12) provides a convenient, approximate estimation of J in the elastic, small-scale yielding and fully plastic regimes.

2.3   The R6 Failure Assessment Diagram Method

With J-estimation schemes, fracture is conceded when J reaches a critical material value, J_{mat}. This may be written in terms of an equivalent fracture toughness, K_{mat}:

$$J_{mat} = K_{mat}^2/E^1 \qquad (13)$$

where

$$E^1 = E/(1-v^2) \qquad (14)$$

in plane strain and $E^1 = E$ in plane stress. Similarly G can be written in terms of the elastic stress intensity factor K as

$$G = K^2/E^1 \qquad (15)$$

Within the R6 method [8], fracture is not assessed directly in terms of J but in terms of two parameters. The first of these is L_r, a measure of proximity to plastic collapse:

$$L_r = P/P_{Lc} = \sigma_{ref}/\sigma_Y \qquad (16)$$

The second parameter is K_r, a measure of proximity to linear elastic fracture:

$$K_r = K^p/K_{mat} \qquad (17)$$

Here a superscript p has been added to the stress intensity factor to indicate that it is a value for the primary load P. In view of equations (12-17), avoidance of fracture by $J \leq J_{mat}$ can be written in the equivalent form

$$K_r \leq f_2(L_r) \qquad (18)$$

where $f_2(L_r) = [E \, \epsilon_{ref}/\sigma_{ref} + \frac{1}{2} L_r^2/(E \, \epsilon_{ref}/\sigma_{ref})]^{-\frac{1}{2}}$ (19)

is termed the option 2 failure assessment curve in R6. The R6 approach is depicted in Figure 5. The two parameters L_r and K_r of equations (16, 17) are evaluated, plotted on the diagram as the point (L_r, K_r); failure is then avoided if the point lies within the failure assessment curve. Conversely, failure is conceded if the point lies outside the curve.

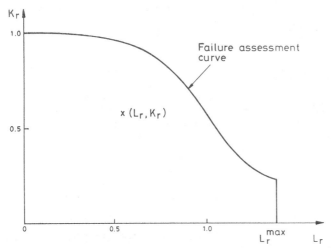

Figure 5 The R6 Failure Assessment Diagram

Figure 5 also contains an upper limit, L_r^{max}, to the value of L_r. This is defined as

$$L_r^{max} = \sigma_f/\sigma_Y \tag{20}$$

where σ_f is a flow stress usually defined as the mean of the yield stress and the ultimate tensile stress.   This provides an explicit limit against plastic collapse, with some allowance for strain hardening beyond yield.

For austenitic stainless steels, which show a large amount of strain hardening beyond yield, equation (19) has been plotted for stress-strain curves measured over a wide range of temperatures [9]. The resulting curve has been found to be sensibly independent of temperature and well described by

$$f_1(L_r) = \{1 - 0.14L_r^2\} \{0.3 + 0.7\exp\ (-0.65L_r^6)\} \tag{21}$$

which is termed the option 1 failure assessment curve in R6. This is a convenient curve for austenitic steels but may also be used as a general curve as it has been found to be a lower bound to option 2 curves for a range of materials. It has the advantage that stress-strain data are no longer required.

R6 also contains an option 3 failure assessment curve

$$f_3(L_r) = (G/J)^{½} \tag{22}$$

which can be used when a detailed solution for J is available. Clearly the option 3 curve is dependent on geometry, crack size and material behaviour which all influence the calculation of J. The option 2 curve which depends only on material behaviour and the option 1 curve which is independent of material and geometry have, however, been found to provide conservative assessments when compared to detailed calculations and experiments on large-scale structures [10].

It is worth noting that R6 has been developed to address a wide range of issues relevant to defective components. A description of these is beyond the scope of the present article but one development which is used later in Section 7 in a worked example is the treatment of combined primary and secondary stresses. Secondary stresses are defined as those which do not affect plastic collapse and may be thermal or welding residual stresses, for example. As they do not affect plastic collapse, they do not affect the parameter L_r. However, they do have a significant effect in the elastic regime and the definition of K_r is, therefore, extended from equation (17) for combined primary and secondary loadings to

$$K_r = (K^p + K^s) / K_{mat} + \rho \qquad (23)$$

Here K^s is the stress intensity factor for the secondary stresses and ρ is a parameter which covers plasticity interactions between primary and secondary stresses. The value of ρ is generally small ($\ll 1$) and methods for its evaluation are given in R6 [8].

3.   ASSESSMENT OF CREEP CRACK INITIATION

3.1   Numerical and Reference Stress Estimates of the Steady State Creep Parameter C*

The parameter C* is the creep equivalent of the J-integral and characterises the stress and strain rate fields near the crack tip in steady state creep. For power-law creeping materials in which creep strain rate $\dot{\varepsilon}^c$ is related to stress by

$$\dot{\varepsilon}^c = \dot{\varepsilon}_o \, (\sigma / \sigma_o)^n \qquad (24)$$

an analogy between power-law plasticity and power law creep enables C* to be written as

$$C* = \sigma_o \, \dot{\varepsilon}_o \, c \, h_1 \, (P/P_o)^{n+1} \qquad (25)$$

in a similar manner to equation (2) for J. Here, P_o is a normalising load proportional to σ_o. If P_o/σ_o is equal to P_Y/σ_Y then h_1 in equation (25) is identical to h_1 in equation (2) for n=N.

The analogy between J and C* also enables the reference stress estimate of J in equation (11) to be used to provide a reference stress estimate of C* as

$$C* = E \ G \ \dot{\varepsilon}^c_{ref}/\sigma_{ref} \qquad (26)$$

where σ_{ref} is again defined by equation (10) and $\dot{\varepsilon}_{ref}^{\ c}$ is the creep strain rate at this reference stress. Equation (26) may be written in an alternative form

$$C* = \sigma_{ref} \ \dot{\varepsilon}^c_{ref} \ R^1 \qquad (27)$$

where

$$R^1 = (K^p/\sigma_{ref})^2 \qquad (28)$$

This form is used in R5 [2] and contains a small conservatism compared to equation (26) for plane strain as E^1/E has been set equal to unity.

As both K^p and σ_{ref} are proportional to applied load, R^1 is a characteristic dimension independent of load. Solutions for R^1 for a centre cracked tension geometry and a single edge notched geometry under tension and bending are shown in Figure 6. For small crack sizes, R^1 is proportional to crack size; for deep cracks R^1 is proportional to remaining ligament size (w-a); at intermediate crack sizes R^1 is typically equal to half the section width w.

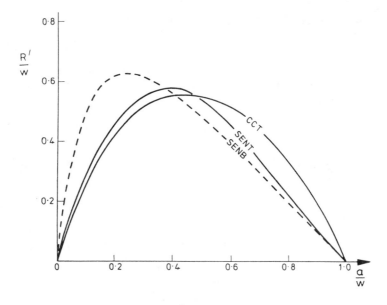

Figure 6 Characteristic Dimension R^1 of Equation (28) for Various Geometries

In view of the discussion in Section 2.2, equations (26) and (27) are expected to provide conservative overestimation of C*. This is illustrated in Figure 7 where results of equation (27), denoted C^*_{ref}, are compared to values of C*, denoted C^*_{exp}, which have been deduced from experimental displacement measurements on compact tension specimens. The evaluation of C^*_{exp} is described elsewhere in this volume in the Chapter by Nikbin.

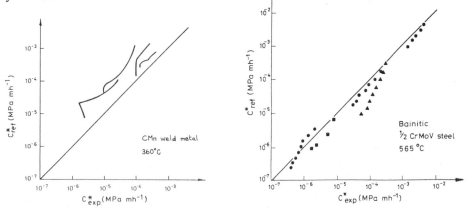

Figure 7 Comparison of Experimental and Reference Stress Estimates of C*

It is worth remarking that the creep response of the materials in Figure 7 could not be adequately described by a power law. However, equation (27) does not require such a representation but enables creep laws including primary, secondary and tertiary parts to be used or raw creep data to be used directly if an equation fitting the data is not available. Equation (27) also allows creep strain accumulation under rising stress, as a crack grows, to be treated by a strain hardening rule. Such a rule is shown schematically in Figure 8 and has been used to evaluate C^*_{ref} in Figure 7.

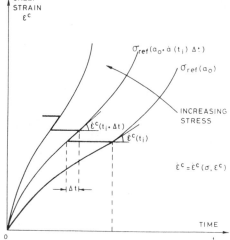

Figure 8 Schematic Strain Hardening Rule

3.2   Transient Creep

Transient creep effects which occur prior to steady state creep are briefly discussed in this sub-section. More detail may be found in Section 4.5.2 of Webster and Ainsworth [1]. A parameter C(t) now describes the crack tip fields rather than the steady state creep parameter C*. For a crack in a component which responds essentially elastically upon loading at time t=0,

$$C(t) = K^2/(n+1) E^1 t \quad , \quad t \to 0 \tag{29}$$

where K is the total stress intensity factor, $K = K^p + K^s$, for the loadings at t=0. At long times

$$C(t) = C* \quad , \quad t \to \infty \tag{30}$$

An estimate of the transition time, t_T, between these two extremes may be obtained by assuming that C(t) reduces according to equation (29) until it reaches C*. Then

$$t_T = K^2/(n+1) E^1 C* \tag{31}$$

In practice, numerical results [1] show that the value of C(t) reduces more slowly for t > 0 than predicted by equation (29) and a better estimate of redistribution time is

$$t_{red} = K^2/E^1 C* \tag{32}$$

For components loaded purely by primary loading, the reference stress estimate of C* in equation (26) may be combined with equations (15) and (32) to give

$$t_{red} = \sigma_{ref}/E \, \dot{\epsilon}^c_{ref} \tag{33}$$

or, more generally for materials which do not exhibit a constant, secondary creep strain rate

$$\epsilon^c(\sigma_{ref}, t_{red}) = \sigma_{ref}/E \tag{34}$$

That is, the accumulated creep strain at the reference stress level at the redistribution time is equal to the elastic strain at the reference stress. This is also the timescale which governs redistribution to the steady state in components without defects [1, 11]. Therefore, at the time t_{red}, redistribution to steady state creep conditions is essentially complete both at the crack tip and in the bulk of a defective component.

As C(t) exceeds C* during the redistribution time, the stress and strain rate fields at the crack tip are higher than those in the steady state during this period. In practice, for components which operate for times in excess of the redistribution time, the integrated effect of the transient field is of interest rather than the magnitude of C(t) itself. Since crack tip events are largely strain controlled and strain rates are proportional to $[C(t)]^{n/(n+1)}$, the integrated effect of interest is

$$\int_{o}^{t} [C(t)]^{n/(n+1)} \, dt = C*^{n/(n+1)} \, t \, [1+\sigma_{ref}/\epsilon^{c}(\sigma_{ref}, \ t)] \qquad (35)$$

where details of the approximations leading to the result are given in [1]. Equation (35) shows that the integrated effect is equal to a factor times that obtained by assuming that steady creep conditions apply at all times. The factor is close to unity provided the accumulated creep strain at the reference stress is much larger than the elastic strain at the reference stress. Equation (35) shows that at the redistribution time of equation (34), the factor equals 2. This factor of 2 is used later in Section 4 to allow for potentially increased crack growth rates during the redistribution period. The more general result of equation (35) is used in sub-section 3.3 below in the estimation of initiation time.

3.3   Estimation of Initiation Time

The incubation time shown schematically in Figure 1 is illustrated by test specimen results in Figure 9. During the initiation time, large strains occur near the crack tip. In the absence of crack extension these strains blunt an initially sharp crack into a rounded notch as shown schematically in Figure 1 and in more detail in Figure 10.

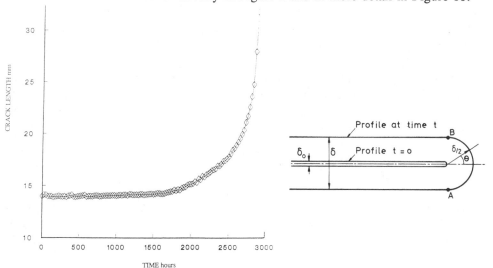

Figure 9   Test Data Showing Extended Initiation Time Prior to Crack Growth

Figure 10   Schematic of Blunted Crack Tip

Analysis of the blunting process is described in detail in Webster and Ainsworth [1] and, therefore, only the essential steps are summarised here. First, from dimensional arguments, the creep strain rate, $\dot{\epsilon}^c_{notch}$, near the notch surface AB in Figure 10 must be of the form

$$\dot{\epsilon}^c_{notch} \propto (\dot{\delta}/\delta) \tag{36}$$

where $\dot{\delta}$ is the rate of notch opening. Secondly, for the creep law of equation (24), the corresponding stress near the notch surface, σ_{notch}, follows as

$$\sigma_{notch} \propto \sigma_o (\dot{\delta}/\dot{\epsilon}_o \delta)^{1/n} \tag{37}$$

Thirdly, as C(t) has the same dimension as C* in equation (27), it is proportional to the product of notch stress, notch strain rate and a length scale; the relevant length scale is the notch opening δ rather than R^1 in equation (27) so that equations (36) and (37) lead to

$$C(t) \propto \sigma_o \, \dot{\epsilon}_o^{-1/n} \, \delta \, (\dot{\delta}/\delta)^{(n+1)/n} \tag{38}$$

Rearranging this expression and integrating leads to the crack opening displacement at time t as

$$\delta^{n/(n+1)} = (\dot{\epsilon}_o/\sigma_o^n)^{1/(n+1)} \int_o^t [C(t)]^{n/(n+1)} dt \tag{39}$$

where the initial crack opening displacement (COD) has been neglected and it transpires from known distributions of strain near blunt notches that the final integration constant in equation (39) is approximately unity.

The integral in equation (39) has already been evaluated in equation (35) and, therefore, equation (39) leads to the following results. First, for cases where the initiation time, t_i, is well in excess of the redistribution time, a constant COD at initiation is equivalent to a correlation between t_i and C* of the form

$$t_i \propto C*^{-n/(n+1)} \tag{40}$$

for secondary creep with a constant C*. Some experimental data showing this trend are plotted in Figure 11. In practice such data can be described by a best-fit or bounding line to the data rather than assuming that the exponent in equation (40) is equal to $-n/(n+1)$.

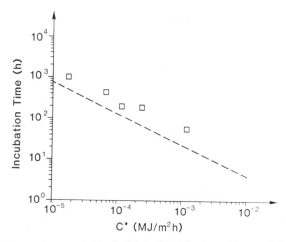

Figure 11 Experimental Variation of Initiation Time with C*

Where transient creep effects are important, or the material exhibits primary creep strains, equations (35) and (39) may be combined with the reference stress estimate of C* in equation (27) to relate the initiation COD to a strain ε^c_i where

$$\varepsilon^c_i = (\delta_i / R^1)^{n/(n+1)} - \sigma_{ref}/E \tag{41}$$

The time for initiation t_i, is then obtained by entering the creep curve for the reference stress at this strain level, Figure 12.

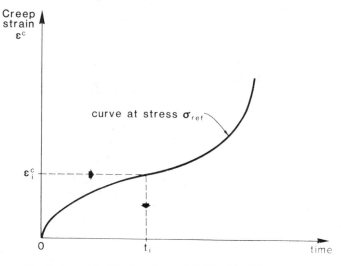

Figure 12 Schematic Evaluation of Initiation Time

If the left-hand-side of equation (41) is less than zero, then creep crack growth should be assumed to start at time t=0. If redistribution is complete, then $\varepsilon^c_i > \sigma_{ref}/E$ and a conservative form of equation (41) is

$$\epsilon^c_i = 0.5 \ (\delta_i/R^1)^{n/(n+1)} \qquad (42)$$

Direct experimental validation for this expression has been demonstrated for its use in the R5 procedures and is reported in [1].

4.   CREEP CRACK GROWTH

Models for creep crack growth and creep crack growth data are discussed in detail in [1] and elsewhere in this volume by Nikbin and so only the essential results are quoted here for use later in this Chapter. Generally, experimental creep crack growth rates are found to correlate with the parameter C* in the form

$$\dot{a} = D_o \ C*^\phi \qquad (43)$$

where D_o and ϕ are material and temperature dependent constants. Models suggest $\phi \sim n/(n+1)$. In the absence of data providing the constants in equation (43), models suggest that

$$\dot{a} = 3C*^{0.85}/\epsilon^*_f \qquad (44)$$

for à in mm/h, C* in MPam/h and ε_f^*, the creep ductility, as a fraction. For plane stress conditions, ε_f^* can be taken as the uniaxial ductility; for plane strain it can be taken as the uniaxial ductility divided by 50 to give a conservative estimate of crack growth rates in equation (44).

Equations (43) and (44) are relevant to steady state creep conditions described by C*. The value of C* may be estimated using the methods described in Section 3.1. However, for times less than the redistribution time, the parameter C(t) controls the magnitude of the crack tip stress and strain rates. In this regime, a pragmatic approach of estimating crack growth rates is to replace C* in equations (43) and (44) by C(t). It should be recognised, however, that experimental data often show reduced crack growth rates at short times so that this is likely to be a conservative approach. As discussed in Section 3.2, the cumulative effect of the transient period is equivalent to a factor of 2 from equation (35) at the redistribution time. Thus, for example, if creep crack growth rates in the steady state are described by equation (43), an estimate of creep crack growth may be made for times greater than the redistribution time by

$$\begin{aligned} \dot{a} &= 2 \ D_o \ C*^\phi \quad , \ t < t_{red} \\ \dot{a} &= D_o \ C*^\phi \quad , \ t > t_{red} \end{aligned} \qquad (45)$$

Once creep crack growth rates are obtained from equations (43-45), then

$$\Delta a = \int \dot{a} \; dt \qquad \qquad \textbf{(46)}$$

In evaluation of C* to obtain à, then the methods of Section 3.1 may be used, noting that the reference stress increases as the crack grows and, therefore, it may be necessary to employ strain hardening rules, Figure 8, to obtain the relevant creep strain rate.

5.   CONTINUUM DAMAGE ASSESSMENT

As noted in Section 1, failure may be governed by continuum damage accumulation in the ligament ahead of the crack rather than by crack growth. Continuum damage describes the formation of microstructural damage in a material under stress in the creep range. Such damage rarely occurs uniformly in structures; the material in the region of stress concentrations becomes heavily damaged first. This local formation of damage tends to weaken the stress concentration through the transfer of stress from damaged material to regions away from the stress concentration feature. This is illustrated in Figure 13 which shows this as a further period of stress redistribution (II) after steady state creep conditions have been established following the stress redistribution after initial loading (I). It can be seen from Figure 13 that as regions away from the initial peak stress pick up stress, these regions then accumulate damage and the processes of damage accumulation and stress transfer continue and spread, or propagate, throughout the structure under failure occurs (III).

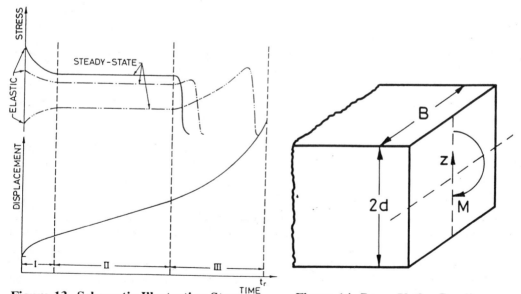

Figure 13 Schematic Illustrating Stress Redistribution Stages in a Structure

Figure 14 Beam Under Bending

The propagation of damage is illustrated first by considering a simple uncracked beam under bending, Figure 14. The analysis is set out in detail in Section 3.5.1 of [1] using a simple ductility measure of damage, Figure 15, and is therefore only briefly presented here. Initiation of damage occurs at the outside of the beam when the accumulated creep strain there becomes equal to the ductility, ϵ_f. The creep strain rate distribution in a beam under bending under steady state creep is

$$\dot{\epsilon}^c = \dot{\epsilon}_o \ [M/Bd^2\sigma_o]^n \ [1 + 1/2n]^n \ (z/d) \qquad (47)$$

for the creep law of equation (24) and the dimensions defined in Figure 14.

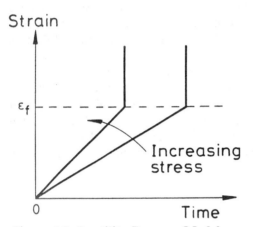

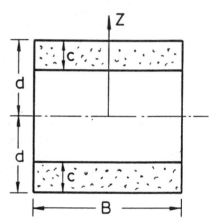

Figure 15 Ductility Damage Model   Figure 16   Beam with Damage Zones

Equating the strain at $z = d$ in equation (47) to the ductility ϵ_f gives the time for damage initiation as

$$t_i = \frac{\epsilon_f}{\dot{\epsilon}_o [M/Bd^2\sigma_o]^n \ [1 + 1/2n]^n} \qquad (48)$$

Following initiation, the outer fibres of the beam can be considered to have 'failed' and no longer carry stress. Provided the material is ductile and can continue to strain, the beam will continue to creep but with the strain rate distribution of equation (47) modified by replacing d with d-c, where c is the depth of the damage zones at the outsides of the beam, Figure 16. The position of the damage front, c, is defined by the strain there being equal to the ductility. As the strain distribution is linear with distance from the neutral axis, the accumulated strain at a small distance δc below the damage front is

$$\epsilon^c/\epsilon_f = (d-c-\delta c)/(d-c) = 1-\delta c/(d-c) \qquad (49)$$

The damage front will propagate the distance δc in a time δt when the creep strain $\dot{\epsilon}^c \delta t$ is sufficient to increase this strain, ϵ^c, to the ductility. Hence the rate of propagation of the damage front is

$$\dot{c} = \dot{\epsilon}^c \, (d-c)/\epsilon_f \tag{50}$$

As the creep strain rate is given by equation (47), with d replaced by (d-c), then

$$\dot{c} = \dot{\epsilon}_o \, [M/B(d-c)^2 \sigma_o]^n \, [1 + 1/2n]^n \, [(d-c)/\epsilon_f] \tag{51}$$

This may be integrated to give the time for growth of the damage front from the outside of the beam (c=0) to the centre (c=d):

$$t_g = \frac{\epsilon_f}{\dot{\epsilon}_o [M/Bd^2\sigma_o]^n \, [1 + 1/2n]^n \, 2n} \tag{52}$$

which is simply

$$t_g = t_i/2n \tag{53}$$

in view of equation (48). The total time for continuum damage failure is $t_{CD} = t_i + t_g$ which follows from equations (48) and (52) as

$$t_{CD} = \frac{\epsilon_f}{\dot{\epsilon}_o \, [M/Bd^2\sigma_o]^n \, [1 + 1/2n]^{n-1}} \tag{54}$$

For this example, this failure time is dominated by initiation of damage. However, in components with higher stress concentrating features the propagation time may be a significant part of the total life.

5.1   Reference Stress Methods for Continuum Damage

The calculations outlined above for the simple beam under bending with a simple damage law illustrate the complexity of analysing the spread of damage in structures. This has prompted the development of simplified methods of life prediction. One of these methods has involved use of the reference stress defined by equation (10) in terms of the limit load. Now, steady state creep solutions generate reasonably uniform stress fields, particularly for high value of the stress index n and later in life when further stress redistribution occurs, Figure 13. Similarly, limit load solutions correspond to regions of uniform stress. Therefore, it transpires that a reasonable estimate of continuum damage failure time is simply

$$t_{CD} \approx t_{r,ref} \tag{55}$$

where $t_{r,ref}$ is the failure time of a uniaxial specimen at the stress level, σ_{ref}.

The estimate of equation (55) may be compared with the result of equation (54) for the beam under bending. For the beam, the limit load in equation (10) is $Bd^2\sigma_Y$ so that

$$\sigma_{ref} = M/Bd^2 \tag{56}$$

For the creep law of equation (24) and the ductility model of Figure 15, the corresponding failure time is

$$t_{r,ref} = \epsilon_f / [\dot{\epsilon}_o (M/Bd^2\sigma_o)^n] \tag{57}$$

Therefore, equations (54) and (57) show that

$$t_{CD}/t_{r,ref} = (1 + 1/2n)^{-(n-1)} \tag{58}$$

which is close to unity for all values of n.

The reference stress methods also work for defective structures. This is shown in Figure 17 where failure times of various test specimens are correlated with the reference stress of equation (10) with P_{Lc} evaluated as the limit load for the crack size at the start of the tests. Although creep crack growth occurred in these tests, it is apparent that the correlation is close to that for uniaxial plain bar specimens confirming the accuracy of equation (55).

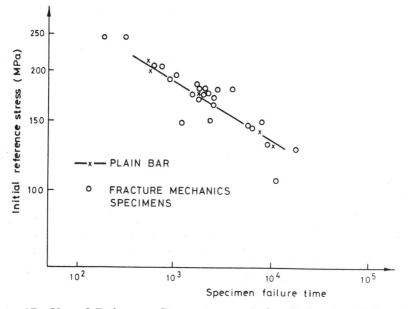

Figure 17   Use of Reference Stress Approach for Estimating Failure Times in Cracked Specimens

That the reference stress method works for cracked structures is surprising at first given the high stress concentration at the crack tip. However, finite-element analysis for a sharp notch, Figures 18 and 19, provides some insight. Figure 19 plots the position of the damage region as a function of time normalised by this life estimate of equation (55). It can be seen that although damage initiates early at the notch, the damage front effectively blunts the notch and a significant amount of time is spent in the damage propagation stage. In this case, failure occurs when the two damage zones meet in the centre of the plate at a time of $0.81t_{r,ref}$. Thus, again equation (55) provides a reasonable estimate of failure time. A similar result would have been obtained for a cracked specimen provided crack propagation follows some way behind propagation of the damage front.

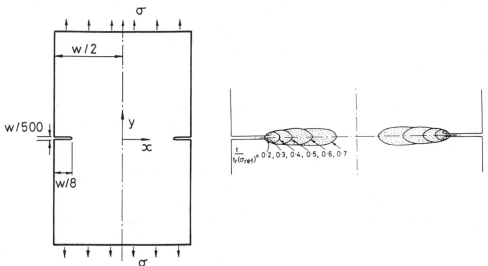

Figure 18 Geometry of Double Edge Notched Specimen

Figure 19 Spread of Damage Fronts Across the Specimen with Time

Overall, therefore, a simple estimate of continuum damage life can be obtained from equation (55). If this does not provide an acceptable life estimate for a cracked component, then more detailed calculations of crack initiation and growth will not increase the life estimate. In a component life assessment procedure, the continuum damage estimate is, therefore, a useful first step and this is addressed next in Section 6.

6.   OVERALL DEFECT ASSESSMENT PROCEDURE

In this section a basic high temperature defect assessment procedure is briefly described. This is then illustrated in Section 7 by means of worked examples. The procedure is based on the Nuclear Electric R5 procedure [2], the British Standards document PD6539 [3] and other procedures in the literature. The overall procedure is described in Section

8.1 of [1] and summarised in the flow chart of Figure 20. The elements in this flow chart are discussed here, with reference back to the methods set out in Sections 2-5.

Prior to performing the defect assessment, it is necessary to establish some basic information on the plant as indicated on the flow chart. Then the initial defect must be characterised, Figure 21. This usually leads to a defect found by inspection being treated as a semi-elliptical surface defect , an embedded elliptical defect or a through-wall defect. Where a defect is postulated, for the purposes of demonstrating that a component is defect tolerant for example, then one of these shapes is also assumed.

The other basic information required in the flow chart is material properties. For example, creep rupture and creep strain data may be represented as illustrated in Figure 22. From such information, it is possible to derive creep strain rate as a function of stress as is required for evaluation of C^* by equation (27).

Once this basic information has been collected, the flow chart of Figure 20 has a box asking whether fatigue is significant. While methods of addressing creep-fatigue crack growth are available [1, 2], a discussion of these in detail is beyond the scope of this chapter. Here attention is concentrated on a creep crack growth assessment. This stage in Figure 20 is amplified in the flow chart of Figure 23.

The defect assessment procedure is described in Section 8.3 of [1] in detail. This procedure addresses the mechanisms shown in Figure 1 which have been described in earlier parts of this chapter. The first step of the defect assessment procedure, Figure 23, is calculation of the margin against time-independent fracture and this may use the failure assessment diagram of R6, Figure 5. Next, assuming this margin is acceptable, the rupture life is calculated for the initial defect size, following the approach in Section 5. This is illustrated in Figure 24 for the creep data of Figure 22. It is necessary to check that this exceeds the sum of the operating period at the time of the assessment, t_o, and the future life required of the plant, t_s. It can been seen from Figure 23 that the time-independent fracture assessment and the rupture assessment are repeated as the crack grows by creep. As these assessments require only a knowledge of the reference stress and the stress intensity factor, see Sections 2.3 and 5, this is straightforward.

The next stage in the flow chart is to determine whether or not the crack grows. Various methods are available for calculating incubation time, Section 3, and if necessary, subsequent creep crack growth, Section 4. These are illustrated in Section 7 by a number of worked examples. It can been from Figure 23 that in performing these calculations, it is necessary to determine whether steady state creep conditions have been established. This is achieved by comparing the time with the redistribution time of equations (33, 34). If the redistribution time is not exceeded, then corrections for non-steady creep must be made using the approaches described in Section 3.3 for incubation and by equation (45) for growth.

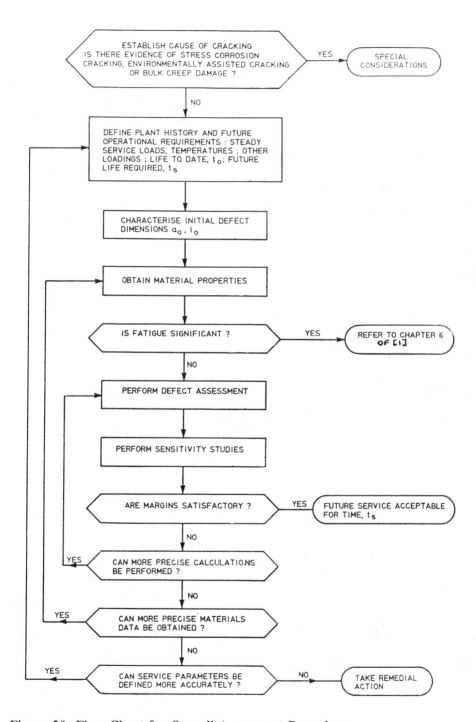

Figure 20 Flow Chart for Overall Assessment Procedure

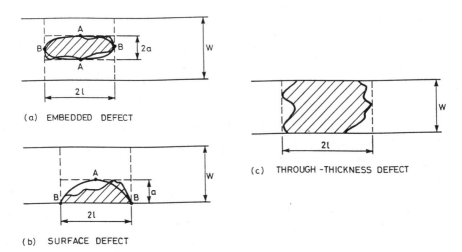

Figure 21 Flaw Characterisation

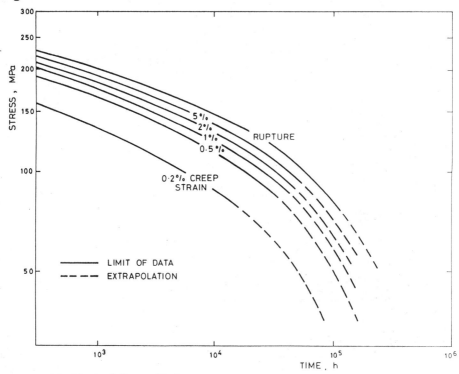

Figure 22 Typical Creep Data

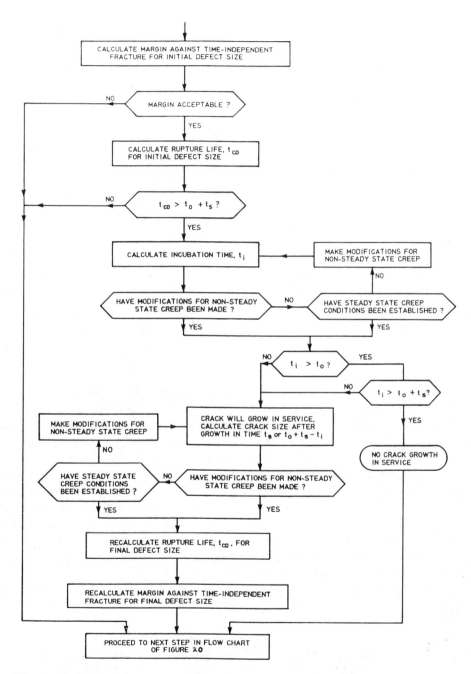

Figure 23 Defect Assessment Flow Chart

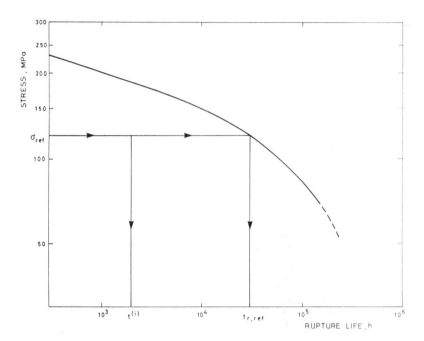

Figure 24 Schematic Creep Rupture Assessment

On completion of the defect assessment part of the procedure, the next stage is a sensitivity study, Figure 20. The principles of sensitivity analysis are discussed in [8]. Essentially, it should be shown that realistic variations in the input parameters do not significantly affect the conclusions of the assessment. If the assessment does not demonstrate acceptable margins, or the sensitivity analysis shows that there is uncertainty, then various options are available as indicated in Figure 20. One of these is to perform more precise calculations by estimating C* from equation (25) or from finite-element analysis instead of from reference stress methods, for example. However, there is often greater uncertainty the material properties used in high temperatures assessments. Therefore, obtaining cast specific rather than general properties or collecting creep crack growth data in the form of equation (44), can be pursued. If it is not possible to demonstrate acceptable margins even after refining the assessment, then remedial action is needed, Figure 20. This may involve a change to the operating conditions, inspection, plant replacement or repair.

Although there are many steps in the procedures set out in the flow charts of Figures 20 and 23, these are relatively easy to follow using the approximate methods set out in Sections 2-5. To illustrate these methods, this chapter concludes by illustrating the procedure by means of a number of worked examples.

7.   WORKED EXAMPLES

Five worked examples are considered. The first three are for an edge cracked plate with differing complexity to illustrate the various methods. The geometry and material properties are first set out in Section 7.1 and the analyses and results are then summarised in Section 7.2. The fourth example, described in Section 7.3, is a simple pressure vessel under constant internal pressure and is used to illustrate the sensitivity of the failure time to some input data. The final example is a surface axial defect in a pressure vessel and this is used to illustrate most aspects of a high temperature component defect assessment. Further details of all examples may be found in [1].

7.1   The Single Edge Cracked Plate and Material Properties

The edge cracked plate being considered is shown in Figure 25. For all the calculations, Young's modulus is taken as 185000 MPa. For the first two sets of calculations only creep rupture and creep crack growth are assessed and simple power law material behaviour is assumed with creep strain rate, related to stress by equation (24) and rupture time related to stress by

$$t_r = A\sigma^{-\nu} \qquad\qquad (59)$$

where A and ν are constants. The values of the constants in equations (24) and (59) are $\dot{\varepsilon}_o = 6.25.10^{-5}\text{h}^{-5}$, $\sigma_o = 400\text{MPa}$, $n = \nu = 16$ and $A = 4000(400)^{16}$ for σ in MPa and t_r in hours. Creep crack growth follows from equation (44) with $\varepsilon^*_f = 0.25$ and $\dot{a}$ in units of mm/h for C* in units of MPa m/h.

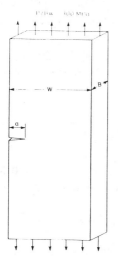

Figure 25 Single Edge Cracked Plate

For the third set of calculations a more complex creep law is followed using a θ-parameter fit to the data in Figure 26. Rupture is assumed to occur when the creep strain reaches 25%. Initiation of creep crack growth is assumed to start when a critical crack opening displacement of $\delta_i = 0.06$mm is reached and subsequent creep crack growth follows equation (43) with $D_o = 6$ and $\phi = 0.85$.

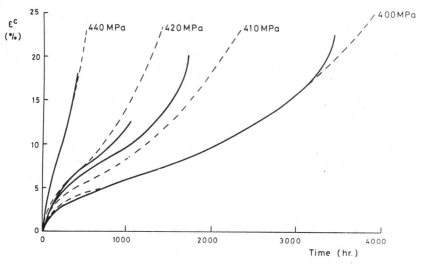

Figure 26 Creep Data for Example 3; θ-parameter Fit Shown by Dashed Lines.

7.2   Calculations for Examples 1-3

For examples 1 and 2, the creep rupture assessment is identical and is based on calculation of a reference stress and then evaluation of the life, t_{CD}, from the corresponding rupture time at this stress, equation (55). For this geometry, the limit load is given by equation (5) and the reference stress therefore follows from equation (10) as

$$\sigma_{ref} = 0.866 \ (P/Bw) \ / \ [1 - a/w - 1.232 (a/w)^2 + (a/w)^3]) \quad \textbf{(60)}$$

which has a value $\sigma_{ref} = 342$ MPa for an applied stress P/Bw = 300 MPa and initial crack size $a_o = 20$mm. Hence, using the creep rupture data of equation (59)

$$t_{CD} = 4000 \ (342/400)^{-16} = 48137h.$$

The only other calculation performed for examples 1 and 2 is creep crack growth, which requires calculation of C^*. For example 1, C^* is calculated using the solutions of [6] for power-law materials. These require some adjustment to the normalising load to enable interpolation of a/w as described in Section 2.1. For a/w = 0.2, the value of h_1 from equation (7) is h_1 = 5.89 using equations (5) and (6). For a normalising stress σ_o = 400MPa, the value of $P_o = P_Y\sigma_o/\sigma_Y$ and P_o/Bw = 363MPa for a/w = 0.2. Then using the constants summarised in Section 7.1, noting that c=a(1-a/w) for this geometry,

$$C^* = 400 \times 6.25 \times 10^{-5} \times 0.02 \times 0.8 \times 5.89 \times \left(\frac{300}{363}\right)^{17}$$

$$= 9 \ 10^{-5} \text{ MPa mh}^{-1}$$

The corresponding creep crack growth rate from equation (44) is

$$\dot{a} = 4.4 \ 10^{-3} \text{ mmh}^{-1}$$

Subsequent crack growth rates are calculated in a similar manner leading to the results in Figure 27. It can be seen that significant creep crack growth occurs in times much shorter than the calculated rupture life. Thus, crack growth rather than rupture is dominant in this case.

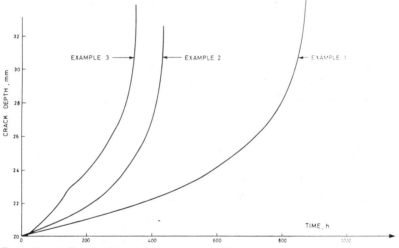

Figure 27 Calculated Creep Crack Growth for Examples 1-3

For example 2, the simplified reference stress methods of equation (27) are used to calculate C^*. For this geometry, the stress intensity factor is

$$K = P\sqrt{2w\tan\psi} \frac{[0.752 + 2.02(a/w) + 0.37(1 - \sin\psi)^3]}{Bw\cos\psi}$$

where $\psi = \pi a/2w$

For a_o = 20mm this gives K = 103 MPa m$^{1/2}$ for the applied stress of P/Bw = 300 MPa. Consequently from equation (28)

$$R^1 = (103/342)^2 \text{ m} = 90\text{mm}$$

using the reference stress of equation (60). The value of C* is obtained from equation (27) using the creep strain rate of equation (24) with the constants detailed in Section 7.1 as

$$C^* = 342 \times 6.25 \times 10^{-5} \times \left(\frac{342}{400}\right)^{16} \times 0.09$$

$$= 1.6 \times 10^{-4} \text{ MPa mh}^{-1}$$

The corresponding creep crack growth rate is

$$\dot{a} = 7.1 \times 10^{-3} mmh^{-1}$$

This shows a factor of about 1.6 on initial creep crack growth rate, compared to that using the power-law solution. Subsequent creep crack growth rates follow by updating K and σ_{ref} as the crack size increases leading to the results in Figure 27.

For example 3, both creep rupture and creep crack incubation are evaluated from the creep strain data of Figure 25. The initial reference stress is given by equation (60). The rupture life for a rupture ductility of 25% follows from the fit to Figure 26 at this stress of 342MPa as

$$t_{CD} = t_{r,ref} = 5.43 \times 10^4 h$$

Incubation is evaluated from equation (41), which for the value of δ_i given in Section 7.1 and the value of R^1 calculated above leads to

$$\varepsilon_i^c = \left(\frac{0.06}{90}\right)^{0.85} - \frac{342}{185000}$$

$$= 0.00014$$

From the creep strain data of Figure 25 at the reference stress of 342 MPa, the incubation time is then

$$t_i = 11.0 h$$

From equation (34), this is less than the redistribution time since the creep strain is less than

$$\sigma_{ref}/E = 0.00185$$

By entering the creep curve at the reference stress at this value of strain, the redistribution time of equation (34) is

$$t_{red} = 144h$$

As incubation occurs before the redistribution time, modifications for transient creep are required in the early stages of crack growth. Therefore, the initial crack growth rate is given from equation (45) and the constants in Section 7.1 as

$$\dot{a} = 12C^{*0.85}$$

At t_i, the creep strain rate from Figure 25 is

$$\dot{\epsilon}^c = 1.3 \times 10^{-5}h^{-1}$$

and the corresponding value of C^* from the reference stress formula of equation (27) is then

$$C^* = 4 \times 10^{-4} \text{ MPa } mh^{-1}$$

so that

$$\dot{a} = 1.6 \times 10^{-2}mmh^{-1}$$

Because of the primary stage in Figure 25, this is a higher strain rate than for examples 1 and 2 and a correspondingly higher crack growth rate. The calculations of crack growth have been performed in an iterative manner as for examples 1 and 2, but with a change in crack growth to

$$\dot{a} = 6C^{*0.85} \text{ for t } \geq t_{red}$$

Additionally a strain hardening rule has been used to evaluate creep strain rate as the crack grows, Figure 8.

The overall crack growth calculations for the three cases are summarised in Figure 27. It can been seen that the inclusion of creep crack incubation and treatment of transient creep complicates the assessment for example 3 but the same trends of accelerating crack growth are obtained in all cases.

7.3   Example 4 - Simple Pressure Vessel

The fourth example is a thick-walled pressure vessel with external radius r_o = 175mm and internal radius r_i = 115mm. The vessel contains a fully circumferential external defect of depth a_o = 30mm. The vessel is made of a normalised and tempered ½ CrMoV steel and operates at a constant pressure of p=62.5MPa at a temperature of 565°C.

Creep data for this material are adequately described by the secondary/tertiary creep law

$$\varepsilon^c = \varepsilon_f \left[1-(1-t/t_r)^{1/\gamma}\right] \tag{61}$$

with

$$t_r = 5 \times 10^{18}\, \sigma^{-7} \tag{62}$$
$$\varepsilon_f = \gamma\, \dot{\varepsilon}_s^c\, t_r \tag{63}$$

γ = 6.4, and a secondary creep law

$$\dot{\varepsilon}_s^c = 5.8 \times 10^{-29}\, \sigma^{10.6} \tag{64}$$

Differentiation of equation (61) leads to the creep strain rate as

$$\dot{\varepsilon}^c = \dot{\varepsilon}_s^c / (1- \varepsilon^c / \varepsilon_f)^{\gamma\, -1} \tag{65}$$

in an explicit strain hardening form.

A limit load solution for this case is given in [1] based on a Tresca yield criterion. For cracks greater than half way through the vessel wall, this leads to the reference stress from equation (10) as

$$\sigma_{ref} = \frac{p}{\ln[(r_o-a)/r_i] + \tfrac{1}{2}[1-r_i^2/(r_o-a)^2]} \tag{66}$$

For the applied pressure of 62.5MPa and the dimensions given above, this leads to

$$\sigma_{ref} = 150\text{MPa}$$

for the initial crack size a = a_o = 30mm. Hence, using equation (55) with the rupture data of equation (62) leads to an estimate of the failure time of the pressure vessel by continuum damage mechanics:

$$t_{CD} = 2957\text{h}$$

To assess initiation and growth, a stress intensity factor solution is needed. One based on finite-element results for the range of crack sizes considered here is given in [1] as

$$K = p\sqrt{\pi a}[\ 1.421 + 1.5(a/w - \tfrac{1}{2}) + 21(a/w - \tfrac{1}{2})^3\ ]/(r_o^2/r_i^2 - 1) \qquad (67)$$

where $w = r_o - r_i$ is the wall thickness. For the initial crack size, combination of equations (66) and (67) leads to the characteristic dimension of equation (28)

$$R^1 = 19mm$$

It is worth remarking that this dimension is comparable to the crack depth.

Once this dimension has been determined, the creep strain at initiation can be evaluated from equation (42) and a value of initiation COD. For this material, this is $\delta_i = 0.11mm$ leading to $\varepsilon_i^c = 0.0045$. This is well in excess of the elastic strain at the reference stress, so that widespread creep conditions are established at initiation. Therefore, it is not necessary to consider the transient creep conditions, discussed in Section 3.2, in either the initiation or subsequent creep crack growth calculations.

Entering the creep curve of equation (61) at the reference stress of 150MPa leads to

$$t_i = 612h$$

for the creep strain to equal ε_i^c. The corresponding creep strain rate from equation (65) is $8.1 \times 10^{-4}h^{-1}$ leading to a value of $C^* = 2.32 \times 10^{-5}MPamh^{-1}$ from equation (27). Creep crack growth rates are assumed to be governed by equation (44) with the creep ductility given by equation (63); note, this ductility is a function of stress level. The initial creep crack growth rate is then $2.74 \times 10^{-3}mmh^{-1}$. Integrating the crack growth calculations from the initial depth of 30mm to a depth of 48mm, at which the reference stress and corresponding crack growth rates are very high, leads to a growth time

$$t_g = 1318h$$

and hence a failure time

$$t_f = t_i + t_g = 1930h$$

It is apparent that this is close to the time for continuum damage failure, $t_{CD} = 2957h$, calculated above. The sensitivity of the calculated life to the initiation assessment has been evaluated by considering values of initiation COD, δ_i, in the range of 0 - 0.5mm. At lower values of δ_i than 0.11mm, crack growth starts at shorter times than 612h, but the initial crack growth rate is lower than that calculated above because of the lower strain rate from equation (65). Consequently, the growth time is greater than 1318h. The converse is true at higher values of δ_i. The overall effect of this is shown in Figure 28: while t_i is very sensitive to δ_i, the calculated failure time is much less sensitive, varying by less than a factor of 2 over the full range of values of δ_i.

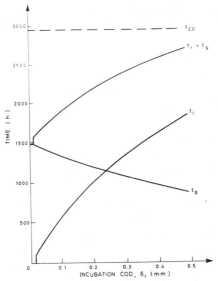

Figure 28 Sensitivity of Calculated Life to Incubation COD.

7.4   Example 5 - Surface Axial Defect in a Pressure Vessel

The final example broadly follows the presentation in [3] and is example 7 in [1]. The
problem is specified as a Type 316 stainless steel cylindrical vessel which operates for
28 days each month. It is shut down at the end of the 28th day and is restarted at the
beginning of the first day of the succeeding month. The monthly operating schedule is
given in the Table below.

OPERATING SCHEDULE		
Month	Pressure (MPa)	Temperature (°C)
April - June, August - October, January - March	4.0	575
November, December	6.0	550
July	Shutdown	

The vessel was put into operation at the beginning of April 1985 and during the July 1990 shut-down a crack was discovered in parent material well away from any welds. There was no evidence that the crack had grown during the first five years of service.

Future operation is required until mid-2005 with the same operating conditions as specified in the Table above.

A defect was located at the outer surface of the vessel, oriented in a axial-radial direction normal to the surface as shown in Figure 29. The defect is characterised as a semi-elliptical surface defect with $a_0 = 7$mm, $l_0 = 20$mm.

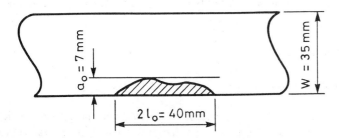

Figure 29 Defect Geometry

The material properties are summarised in the Table below. In some cases, both mean and bounding data are given. 'Worst case' data are assumed in making the assessment.

MATERIAL PROPERTIES		
	550°C	575°C
Young's modulus E (GPa)	149.2	147.2
Lower bound yield stress σ_Y (MPa)	113	112
Lower bound ultimate tensile stress σ_u (MPa)	395	384
Lower bound fracture toughness K_{mat} (MPa m$^{1/2}$)	105	105
Creep strain ε^c (t in h, σ in MPa)	$2.1 \times 10^{-16}\sigma^4 x (t+768t^{1/3})$	$4.3 \times 10^{-16}\sigma^4 x (t + 768t^{1/3})$
Upper bound creep crack growth rate (mm h^{-1})	equation (43) with $D_0 = 23.0$, Ø $= 0.81$	
Mean creep rupture t_r(h) for σ in MPa, T in K	$T(\log_{10}t_r + 22) = 2.6865 \times 10^4 - 25.32\sigma + 1.9748 \times 10^{-2}\sigma^2$	
Lower bound creep rupture	Mean creep rupture life divided by 10	

A check on fatigue shows that this need not be considered [1,3.].

The defect assessment requires a number of calculations. An important element in these is the reference stress which is proportional to pressure, p, by a factor which depends on the crack dimensions. For the initial defect dimensions and a limit load solution given in [1] this leads to

$$\sigma_{ref} = 108.5 MPa \text{ for } p = 6 MPa$$

$$\sigma_{ref} = 72.3 \text{ MPa for } p = 4 \text{ MPa.}$$

The corresponding values of L_r from equation (16) and the yield stress data given above are

$L_r = 0.96$ for high pressure operation

$L_r = 0.65$ for low pressure operation.

For a time dependent assessment it is also necessary to evaluate the parameter K_r of equation (23). The input values of stress intensity factor are listed in the Table below for the pressure loading, K^p, and the thermal stress, K^s. The small plasticity correction factors ρ are also listed with the total values of K_r.

	High Pressure Operation		Low Pressure Operation	
	Deepest Point	Surface	Deepest Point	Surface
K^p (MPa m$^{1/2}$)	13.6	9.0	9.0	6.0
K^s (MPa m$^{1/2}$)	5.7	4.6	5.7	4.6
ρ	0.018	0.02	0.051	0.059
K_r	0.202	0.150	0.192	0.160

In Figure 30, the points (L_r, K_r) are plotted for the most onerous conditions on the failure assessment diagram. the points lie well inside the failure assessment curve demonstrating an adequate margin of safety.

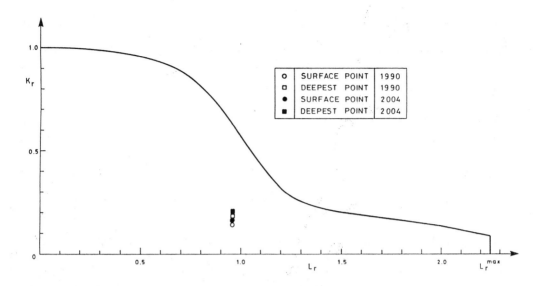

Figure 30 R6 Assessment

As operating pressure varies, the life fraction rule is used for the creep rupture assessment. The plant operated for 28 days (= 672h) each month with 10 months of high pressure and 48 months of low pressure operation between April 1985 and detection of the defect in July 1990. Using creep rupture and the reference stresses given above gives the rupture times, $t_{r,ref}$, of 3.83 x 10^6h for high pressure operation and 4.35 x 10^6h for low pressure operation. Hence,

$$D_c = \frac{10 \times 672}{3.83 \times 10^6} + \frac{48 \times 672}{4.35 \times 10^6} = 0.0092$$

This small value of damage ($D_c < < 1$) corresponds to a large rupture life, t_{CD}, and hence the calculations of crack growth can continue. As the crack is found in service, it is assumed to be growing and no allowance for an initiation period is made. The crack growth calculations are very similar to those in the earlier examples presented, with a strain hardening rule to allow for changes in reference stress due to the changing crack size. As the defect is semi-elliptical, C* has to be calculated for both the surface point and the deepest point.

Some results of performing these calculations are given in the Table below for the month of low pressure operation in August 1990 immediately following detection of the defect and assuming steady state creep conditions to apply.

Crack Growth Calculations for August 1990		
	Deepest point	Surface
C* (MPa mh^{-1})	1.55 x 10^{-8}	6.71 x 10^{-9}
Crack growth rate (mmh^{-1}) from equation (43)	$\dot{a}$ = 1.08 x 10^{-5}	$\dot{l}$ = 5.51 x 10^{-6}
Crack growth (mm) in August 1990	$\Delta a = 2 \times \dot{a} \times 672h$ = 0.0146	$\Delta l = 2 \times \dot{l} \times 672h$ = 0.0074

The thermal stress provides an additional complication in this case when performing the check on steady state creep conditions. Redistribution to the steady state takes longer as the thermal stresses must be relaxed in addition to the elastic stresses due to the mechanical loading. This can be accounted for by using the total stress intensity factor $K = K^p + K^s$ to calculate the redistribution time of equation (32). This is equivalent to modifying equation (34) such that redistribution is complete when

$$\varepsilon^c_{ref} = (\sigma_{ref}/E) \ (K/K^p)^2$$

Inserting the values in the table above into the right-hand side of this equation gives an elastic strain of 1.3 x 10^{-3} in excess of the accumulated creep strain to July 1990 which is calculated as 1.0 x 10^{-3}. Therefore, redistribution is not complete and the crack growth rates must be doubled to evaluate the values as Δa and Δl, during August 1990, see equation (45), as indicated in the Table above The magnitudes of these are sufficiently small that it is not necessary to divide the time steps into less than a month. It is also straightforward to calculate increments of creep damage and creep strain from the reference stress during the month.

The calculations outlined above have been repeated for each successive month. Non-steady creep conditions persist until August 1992. The amounts of crack extension at the deepest point and on the surface are shown in Figure 31 corresponding to a surface defect with a = 9.4mm and 1 = 21.4mm by June 2005. Figure 32 shows the creep damage accumulation up to this time. The time-independent assessment has been repeated for the final defect size and the points are included in Figure 30.

It is clear from all these calculations that the plant can operate to the year 2005 without excessive creep crack growth or failure by creep rupture or fast fracture.

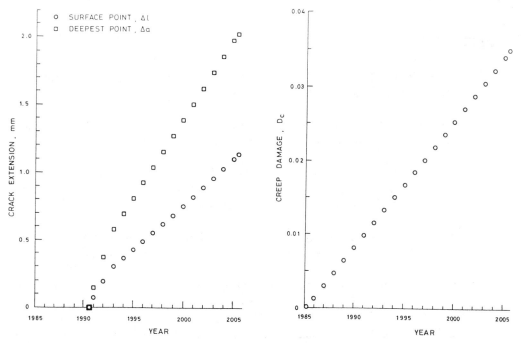

**Figure 31 Crack Growth
Calculations**

**Figure 32 Creep Damage
Calculations**

8   CLOSING REMARKS

In this chapter, the calculations required to follow a basic high temperature defect
assessment procedure have been described in outline. A number of worked examples
have been given to illustrate the procedure and the calculations. References have been
given for those interested in more detail in this area.

Acknowledgement - This Chapter is published with permission of Nuclear Electric Ltd.

REFERENCES

1.   G A Webster and R A Ainsworth, High Temperature Component Life Assessment, Chapman and Hall, London (1994).

2.   Assessment Procedure for the High Temperature Response of Structures, Nuclear Electric Procedure R5 Issue 2, Nuclear Electric Ltd, Gloucester (1997).

3.   Methods for the Assessment of the Influence of Crack Growth on the Significance of Defects in Components Operating at High Temperatures, PD6539, British Standards Institution, London (1995).

4.   V Kumar, M D German and C F Shih, An Engineering Approach for Elastic-Plastic Fracture Analysis, EPRI Report NP-1931, EPRI, Palo Alto (1981).

5.   R A Ainsworth, The Assessment of Defects in Structures of Strain Hardening Material, Engng Fract Mech 19, 633-642 (1984).

6.   C F Shih and A Needleman, Fully Plastic Crack Problems, Part 1: Solutions by a Penalty Method, J Appl Mech 51, 48-56 (1984).

7.   A G Miller and R A Ainsworth, Consistency of Numerical Results for Power-Law Hardening Materials and the Accuracy of the Reference Stress Approximation for J, Engng Fract Mech 32, 233-247 (1987).

8.   Assessment of the Integrity of Structures Containing Defects, Nuclear Electric Procedure R6 Revision 3, Nuclear Electric Ltd, Gloucester (1997).

9.   R A Ainsworth, Failure Assessment Diagrams for Use in R6 Assessments for Austenitic Components, Int J Pressure Vessels Piping 65, 303-309 (1996).

10.   I Milne, R A Ainsworth, A R Dowling and A T Stewart, Background to and Validation of CEGB Report R/H/R6 Revision 3, Int J Pressure Vessels Piping, 32, 105-196 (1988).

11.   R K Penny and D L Marriott, Design for Creep, Second Edition, Chapman and Hall, London (1995).

THE FRACTURE MECHANICS CONCEPTS OF CREEP AND CREEP/FATIGUE CRACK GROWTH

K.M. Nikbin
Imperial College of Science, Technology and Medicine, London, UK

Abstract

This Chapter covers topics related to the creep and fracture of engineering materials at high temperatures containing defects and their relation to high temperature life assessment methods. Following a description of engineering creep parameters basic elasto-plastic fracture mechanics methods are presented and high temperature fracture mechanics parameters are derived from pasticity concepts. Techniques are shown for determining the creep fracture mechanics parameter C* using experimental crack growth data, collapse loads and reference stress. Models for predicting creep crack initiation and growth in terms of C* and the creep uniaxial ductility are developed. These ideas are then applied to practical techniques for analysing and predicting crack initiation and growth under static and cyclic loading conditions. The subject cannot covered in detail within one chapter but rather an overview is presented in order to highlight the relevant topics that the reader should seek further reading on.

Contents

1 Introduction

Engineering life assessment and component design utilise models based on theoretical principles which always need to be validated under practical and operational circumstances. In this chapter engineering creep parameters ranging from uni-axial to multiaxial states of stress are described. The mechanism of time dependent defomrmation is shown to be analogous to deformation due to plasticity. Therefore elasto-plastic fracture mechanics methods are reviewed and linked to high temperature fracture mechanics parameters. Techniques are shown for determining the creep fracture mechanics parameter C* using experimental crack growth data, collapse loads and reference stress methods. Models for predicting creep crack initiation and growth in terms of C* and the creep uniaxial ductility are developed. Cumulative damage concepts are used for predicting crack growth under static and cyclic loading conditions. The topics discussed in this chapter complement the book "High temperature component life assessment" by Webster and Ainsworth [1] in which emphasis is placed on analysis of creep initiation and crack growth in terms of fracture mechanics concepts.

1.1 Creep analysis of uncracked bodies

The time dependent deformation mechanism occuring at elevated temperature that is generally non-reversible is defined as creep. Creep is most likely to occur in components that are subjected to high loads at elevated temperatures for extended periods of time. Creep may ultimately cause fracture or assist in developing a crack in components subjected to stresses at high temperatures. In the last 30 years rapid development has taken place in the subject and references at the end of the chapter give an indication of this work [1-29].

The phenomenon of creep is based on a time dependent process whereby the material deforms irreversably. Creep in polycrystalline materials occurs as a result of the motion of dislocations within grains, grain boundary sliding and diffusion processes. A creep curve can simply be split up into three main sections as shown in figure 1. All the stages of creep are not necessarily exhibited by a particular material for given testing conditions. In figure 1 a region of accelerating creep rate is shown which occurs immediately after the full load has been applied. It is termed an incubation period because it happens prior to the attainment of normal primary, secondary and tertiary creep. It is usually only observed in single crystal or highly oriented materials at relatively low stresses and high temperatures and is associated with a gradual build up of mobile dislocations.

For most cases the first region is the primary creep stage with no initial incubation period observed. This is a period of decreasing creep rate where work-hardening processes dominate and cause dislocation motion to be inhibited. The secondary or steady-state region of creep deformation is frequently the longest portion and corresponds with a period of constant creep rate where there is a balance between work-hardening and thermally activated recovery (softening) processes. The final stage is termed the tertiary region. This is a period of accelerating creep rate which culminates in fracture. It can be caused by a number of factors which include; increase in stress in a constant load test, formation of a neck (which also results in an increase in stress locally), voiding and/or cracking and overageing (metallurgical instability in alloys).

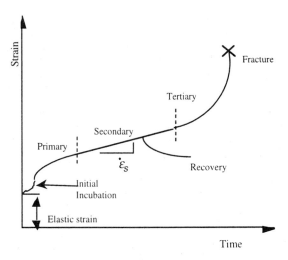

Figure 1: Various regions in a creep curve

A small amount of creep strain is usually recovered, as indicated in figure 1, if the load is removed part way through a creep test. However, for the most part, creep strain can be regarded as permanent. Also, in the absence of voiding creep deformation takes place at constant volume like plastic deformation.

Families of creep curves obtained at different stresses and a constant temperature for most polycrystalline materials will tend to look the same when normalised with time. Similar behaviour is observed when tests are performed at a constant stress and different temperatures. However the primary, secondary and tertiary ratios and the creep ductility could vary both with load and temperature. All stages of creep are accelerated with an increase in stress and/or temperature. The creep ductilities of different materials can be relatively independent, or sensitive, to the testing conditions. Some engineering alloys exhibit an appreciable decrease in failure strain with increase in rupture life.

Generally, the creep properties of crystalline materials can be related to their homologous temperature T_m (i.e. their melting temperature in degrees Absolute). The shapes of creep curve shown in Figure 1 is most relevant to a temperature T greater than about approximately $0.5\,T_m$ (T may range between about 0.4 and 0.6 for different alloys). For tests lasting approximately the same length of time at temperatures above and below this temperature, different shapes will tend to be observed. For $T>0.5T_m$ it is likely that secondary and tertiary creep will be most pronounced and for $T<0.3T_m$ that primary creep will predominate. This behaviour can be partly attributed to the different stresses needed at each temperature to cause failure in the same time. At $T>0.5T_m$ realistic rupture lives are obtained at a stress σ which is less than the yield stress σ_y of a material, whereas at $T<0.3T_m$ a stress of greater than the yield stress would be needed to give the same lifetime. For most materials the primary component of creep tends to be magnified with increase in stress thus causing this term to dominate more at lower temperatures.

Table 1 lists some values of $0.5\,T_m$ (in °C) for a selection of metals. It is apparent that lead is expected to exhibit significant creep at stresses less than its yield stress at room temperature. For the other materials, a stress greater than σ_y would need to be applied to cause significant creep at room temperature. Consequently, for most engineering materials plastic deformation occurs at lower stresses than are needed to cause creep at room temperature whereas at $T > 0.5\ T_m$, approximately, creep occurs at lower stresses than are needed to cause plastic deformation. This observation explains why room temperature design philosophies are based on avoiding yielding and high temperature design codes on avoiding creep failure.

Table 1: Melting temperatures of metals

Material	Melting Temp. °C	0.5 T_m(°C)
Lead	327	27
Aluminium	660	194
Copper	1083	405
Titanium	1690	708
Iron	1530	629
Nickel	1453	590

Creep in polycrystalline materials is sensitive to grain size, alloying additions, initial condition of the material, heat treatment and testing conditions. Figure 2 compares the creep properties of a range of engineering alloys. It indicates the improvements in creep strength that can be gained by alloying additions to steel and why nickel base alloys are used for the highest temperature applications in for example gas turbine engines. The highly alloyed nickel base alloys (often called superalloys) can be used at these high temperatures because they exhibit high creep strengths at relatively high fractions of their homologous temperatures.

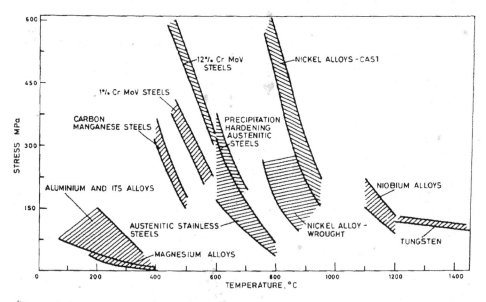

Figure 2: Stress to produce creep rupture in 100 hours

A creep fracture can be transgranular or intergranular [7,16-17] There is a general trend towards transgranular failures at short creep lives and relatively low temperatures and intergranular failures at long lifetimes and higher temperatures. Intergranular failures are usually most relevant to practical operating situations. They can result in creep ductilities, that are much less than room temperature ductilities. Microcracking and voids can be distributed throughout a material at failure or be localized to the final fracture. Several processes contribute to creep deformation in metallic materials. These processes give rise to different stress and temperature dependencies.

1.2 Physical models describing creep

A number of processes dominate the creep processes [1-10] as shown in figure 1. When secondary creep dominates, it is often possible to express secondary creep strain rate $\dot{\varepsilon}_s^c$, in the form

$$\dot{\varepsilon}_s^c \alpha \sigma^n \exp(-Q/RT) \qquad (1)$$

where n and Q are material dependent parameters and R is Boltzmann's constant. The values of n and the activation energy Q are sensitive to the processes controlling creep. The creep processes can be described in terms of deformation mechanism maps which exhibit two main creep fields. In one, creep rate is governed by the glide and climb of dislocations and has a power law stress dependence. In the other, creep is controlled by the stress directed diffusional flow of atoms [9]. For a given mechanism actual creep rates are dependent on material composition, microstructure and grain size. The largest grain size dependence is observed in the diffusional flow region with an increase in grain size resulting in a decrease in creep rate. Solid solution and precipitation hardening alloying additions can impede dislocation motion and influence diffusion rates. There is a general tendency for alloying additions to move mechanism boundaries to higher σ/E and T/T_m ratios [9].

The existence of several creep processes indicates that in general n and Q in equation (1) will change as a mechanism boundary is crossed. Also it has been established that a simple power law stress dependence is not always satisfactory. At high stresses an exponential expression of the form

$$\dot{\varepsilon}_s^c \alpha \exp(\beta\sigma) \exp(-Q/RT) \qquad (2)$$

where β is a material dependent constant is often more adequate. Garofalo [4] has shown that equations (1) and (2) can both be encompassed by the relation

$$\dot{\varepsilon}_s^c \alpha (\sinh \alpha\sigma)^n \exp(-Q/RT) \qquad (3)$$

provided $\beta = \alpha n$. When $\alpha \sigma < 0.8$, equation (3) reduces to equation (1) and when $\alpha \sigma > 1.2$ it reverts to equation (2). No satisfactory physical model has yet been developed which produces expressions of the form of equations (2) or (3).

There are creep laws which deal with the time dependence of creep. The stress and temperature dependence of the material parameters introduced will not be examined. However, for the most part they can be described by expressions that are similar to those used for secondary creep in the previous section. Model based laws [2,12,13] where creep strain is predicted from motion of dislocations give an understanding of the creep behaviour but are rarely useful for engineering purposes.

As a result, empirical laws have been produced [4,5,15] to give more accurate descriptions of the observed shapes of creep curves. A representative selection is listed below with an indication of their ranges of applicability.

Usually for $T/T_m < 0.3$, work hardening processes dominate and primary creep is observed which can often be described by a logarithmic expression of the form

$$\varepsilon^c = \alpha \ln(1 + \beta t)$$

(4)

where α and β are parameters which in general are functions of stress and temperature. Within the temperature range $0.3 < T/T_m < 0.5$, secondary creep begins to appear. A typical equation [5] is

$$\varepsilon^c = \alpha t^m + \dot{\varepsilon}_s^c t$$

(5)

where $m < 1$ and takes the value $1/3$ in the Andrade [2] expression. Again α is in general a function of stress and temperature. In equation (5) the first term describes the primary region and the second term describes secondary creep.

For $T/T_m > 0.5$, equation (5) can still be employed [4] but an alternative expression that has been used [4] is

$$\varepsilon^c = \varepsilon_t \{1 - exp(-t/\tau)\} + \dot{\varepsilon}_s^c t$$

(6)

Other empirical laws have been proposed that have wider applicability than those just presented and which can also accommodate tertiary creep [4,15] Two representative equations are;

$$\varepsilon^c = \alpha t^{1/3} + \beta t + \gamma t^3$$

(7)

and

$$\varepsilon^c = \theta_1 \{1 - exp(-\theta_2 t)\} + \theta_3 \{exp(\theta_4 t) - 1\}$$

(8)

where α, β, γ, θ_1, θ_2, θ_3 and θ_4 are stress and temperature dependent material parameters. In their most general formulations, each of these parameters consists of a summation of terms involving stress and temperature. The equations describe primary, secondary and tertiary creep. They can correlate a wide spread of behaviour because of the number of disposable parameters that are used. They do however need a large body of data to identify all the terms. They are of most use in extrapolating experimental data to longer times.

The list of creep laws presented for describing the time dependence of creep is by no means extensive. The references given will cover the range in depth.

1.3 Time-Temperature Creep Parameters

Design lifetimes of engineering components are often based on time to a specific strain or rupture. For this purpose, it is usually more convenient to replot the type of data shown in Figure 1 as shown in Figs 3 and 4. Secondary creep rates, time to a specific strain (say 1%) or rupture lives can then be read off at any desired stress and temperature. Figure 3 is most appropriate for extrapolating to other stresses at a given temperature and Figure 4 for extrapolating to other temperatures at a particular stress. Straight lines, with the slopes shown, will be obtained on the figures when the data can be correlated by the stress and temperature terms given in equation (1).

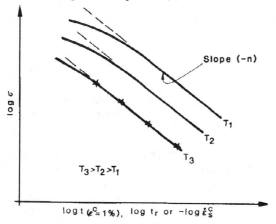

Figure 3: Stress dependence of creep properties at different temperatures.

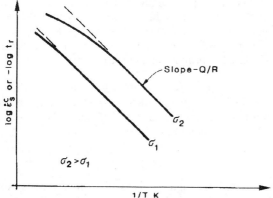

Figure 4: Temperature dependence of creep properties at different at different stresses.

A simple straight line relation is not usually obtained on plots like those depicted in figures 3 and 4 when attempts are made to include a wide spread of data. This is because different values of n and Q are needed to describe different creep mechanisms. Consequently in order to achieve more reliable extrapolations, time-temperature creep parameters have been devised for superimposing all the results onto one so called 'master curve' for one material. The basis of the creep parameters is that time and temperature have similar effects; i.e. the same creep behaviour is obtained at the same stress in a short time at high temperature as is attained in a long time at low temperature. This is clearly a simplification but it does result in satisfactory extrapolations of creep data.

The Sherby-Dorn parameter [2] make use of these type of curves where specific creep data is available. This parameter can be obtained immediately from the temperature dependence shown in equation (1). This equation suggests that creep strain can be written as

$$\varepsilon^c = f(\sigma, \theta) \tag{9}$$

where the stress dependence need not be restricted to a power law function and θ is the Sherby-Dorn parameter

$$\theta = t\ exp\ (-Q/RT) \tag{10}$$

When this relation is valid, creep data obtained at the same stress but different temperatures should superimpose when plotted against θ. Satisfactory agreement is usually found for pure metals and dilute alloys at $T/T_m > 0.5$ but not in other circumstances.

Another relationship the Monkman-Grant [19] is effectively a critical strain criterion. It states that the strain accumulated during secondary creep is a constant at failure so that the product of the secondary creep rate and the rupture life t_r is a constant; i.e.,

$$\dot{\varepsilon}^c_s t_r = C_{MG} \tag{11}$$

where C_{MG} is the Monkman-Grant constant. When secondary creep dominates it predicts a constant creep ductility independent of stress and temperature. This implies that, at a given stress, failure also occurs at a constant value of the Sherby-Dorn parameter. When failure occurs at a finite reduction in area due to internal voiding and cracking $C_{MG}<1/n$ approximately. However, in practice measured creep failure strain may be appreciably greater than C_{MG} due to strain accumulated in primary and tertiary creep.

The widely used Larson-Miller parameter can be expressed in terms of time to a specific strain or time to rupture. Unlike the Sherby-Dorn parameter, which assumes a constant activation energy, this relation implies that activation energy is dependent on stress. The Larson-Miller parameter, P, is usually written in the form

$$P= T\ [C+log_{10}t] \tag{12}$$

where P is a function of stress only, C is a constant and t is the time to a specific strain or the time to rupture. When rupture life is expressed in hours, C is usually in the range 17 to 23 and quite often can be approximated to 20.

An example of a Larson-Miller plot for the nickel base alloy Nimonic 105 is shown in Figure 5, where T is in $°K$ and t_r is the rupture life in hours. Other creep parameters that are cited elsewhere [3] can be used in a similar fashion.

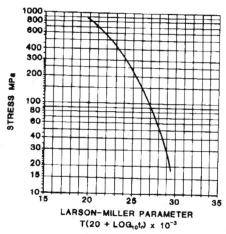

Figure 5: Larson-Miller plot for Nickel-based alloy Nimonic 105.

Almost all basic high temperature material properties data are usually generated at constant stress and temperature. Most components experience changes in stress and temperature during operation. Consequently some procedure is required for extending the application of constant stress and temperature data to variable operating conditions[20-27].

Use is made of experimental observations to develop criteria. Instead a mechanical equation of state approach is employed. With this concept it is postulated that the instantaneous creep rate is governed by the current 'state' of the material and the current stress and temperature conditions imposed, independent of the previous history of these conditions. Several mathematical formulations have been proposed which each define the 'state' of a material differently. There is experimental support for each of these in particular circumstances. They can all be expressed mathematically as

$$\dot{\varepsilon}^c = f(\sigma, T, s) \tag{13}$$

where s is the term which describes the current 'state' of the material and σ and T are the present values of stress and temperature, respectively. The physical expressions relating creep rate to mobile dislocation velocity and density can be regarded as an equation of state [2,12,13]. With this interpretation, the mobile dislocation density in a material would represent its current 'state'. There is seldom sufficient information available to apply a detailed physical expression and almost invariably empirical mechanical equations of state are employed. Several have been proposed to account for different ways of defining the 'state' of the material. The four most common definitions are;

$$\dot{\varepsilon}^c = f_1(\sigma, T, \varepsilon^c) \tag{14}$$

$$\dot{\varepsilon}^c = f_2(\sigma, T, t) \tag{15}$$

$$\dot{\varepsilon}^c = f_3(\sigma, T, \varepsilon^c/\varepsilon_f) \tag{16}$$

$$\dot{\varepsilon}^c = f_4(\sigma, T, t/t_r) \tag{17}$$

The first equation is referred to as the strain hardening (SH) law since it defines the 'state' of the material in terms of the creep strain incurred; the second is called the time hardening (TH), or age hardening, law as it represents 'state' by the time during which creep has been taking place. The latter two equations are normalised versions of the first two. Equation (16) is termed the strain fraction rule (SF) and equation (17) the life fraction rule (LF). In these expressions ε_f and t_r are creep ductility and rupture life, respectively.

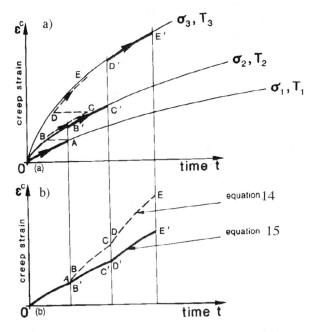

Figure 6: Application of mechanical equations of state to determine a) an instantaneous creep rate change, and b) strain accumulation due to stress (and/or temperature) increases during creep.

An illustration of the application of equations (14) and (15) to stress and/or temperature changes is shown in Figure 6. When the SH law is used a new instantaneous strain rate is obtained after a change in conditions by transferring horizontally at constant strain from one curve to another (i.e. from A to B or from C to D in Figure 6a). With the TH law the transfer is made vertically at constant time (i.e. from A to B' or C' to D'). The strain accumulated due to a progressive change in conditions from σ_1, T_1 to σ_2, T_2 and then to σ_3, T_3 is shown in Figure 6b. For increasingly severe loading conditions, when primary creep behaviour is observed, the strain hardening law predicts more creep strain than the time hardening law. When the conditions become progressively less severe, the reverse is true. During secondary creep all equations of state give the same answer since a horizontal or vertical transfer from one line to another always results in the same creep rate. Where tertiary creep dominates, the SH and TH laws predict the opposite trends to those observed during the primary region. It cannot be argued, therefore, that one of these laws will always give the greatest strain accumulation for all conditions.

The SF and LF rules can be applied in exactly the same way as the SH and TH laws provided the axes on Figure 6 are normalised as $\varepsilon^c/\varepsilon_f$ and t/t_r, respectively. The SF and LF rules are usually more satisfactory than the SH and TH laws for determining final failure under variable operating conditions. With the SF rule, failure is predicted when

$$\Sigma \, \varepsilon^c/\varepsilon_f = 1 \qquad\qquad\qquad (18)$$

and with the LF rule when

$$\Sigma \, t/t_r = 1 \qquad\qquad\qquad (19)$$

where in these summations ε_f and t_r are the creep ductilities and rupture lives relevant to each loading condition. Ideally the equation of state that should be chosen is that which describes the material behaviour best. In most cases insufficient information will be available and judgement has to be employed. When work hardening processes predominate, it may be argued that the SH law (or SF rule) should be most appropriate. When thermal recovery processes control, or metallurgical instability results in ageing, it may be more relevant to choose the TH law (or the LF rule). In any event, the time hardening law will result in the simplest mathematical analysis.

1.4 Complex stress creep

Because creep deformation is not linearly dependent on stress, the effects of stresses that are applied in different directions cannot be superimposed linearly. However, it is found experimentally that;

      (i)   hydrostatic stress does not affect creep deformation;
      (ii) the axes of principal stress and creep strain rate coincide; and
      (iii) no volume change occurs during creep.

These observations are the same as those that are made for plastic deformation [6,20,26]. This is not surprising when both processes are controlled by dislocation motion. The observations imply that the definitions of equivalent stress and strain increment used in classical plasticity theory can be applied to creep provided strain-rates are written in place of the plastic strain increments. Therefore for creep, the Levy-Mises flow rule becomes

$$\frac{\dot{\varepsilon}^c_1}{\left[\sigma_1 - \frac{1}{2}(\sigma_2 + \sigma_3)\right]} = \frac{\dot{\varepsilon}^c_2}{\left[\sigma_i - \frac{1}{2}(\sigma_1 + \sigma_3)\right]} = \frac{\dot{\varepsilon}^c_3}{\left[\sigma_3 - \frac{1}{2}(\sigma_1 + \sigma_2)\right]} = \frac{\dot{\bar{\varepsilon}}^c}{\bar{\sigma}} \qquad (20)$$

where σ_1, σ_2 and σ_3 are the principal stresses, $\dot{\varepsilon}_1$, $\dot{\varepsilon}_2$ and $\dot{\varepsilon}_3$ the respective creep strain rates, $\bar{\sigma}$ is the equivalent stress and $\dot{\bar{\varepsilon}}^c$ corresponding equivalent creep strain rate. Equation (20) satisfies the experimental observations (i) to (iii) provided appropriate definitions are chosen for $\bar{\sigma}$ and $\dot{\bar{\varepsilon}}^c$. From (i), and the observation that dislocations are mainly responsible for creep, it may be inferred that shear stresses govern creep deformation so that either the Von-Mises or Tresca criterion can be employed. With the Von-Mises definition,

$$\bar{\sigma} = \frac{1}{\sqrt{2}}\left[(\sigma_1 - \sigma_2)^2 + (\sigma_2 - \sigma_3)^2 + (\sigma_3 - \sigma_1)^2\right]^{1/2}$$

$$(21)$$

and

$$\bar{\varepsilon}^c = \frac{\sqrt{2}}{3}\left[(\dot{\varepsilon}^c_1 - \dot{\varepsilon}^c_2)^2 + (\dot{\varepsilon}^c_2 - \dot{\varepsilon}^c_3)^2 + (\dot{\varepsilon}^c_3 - \dot{\varepsilon}^c_1)^2\right]^{1/2}$$

$$(22)$$

Assuming the Tresca definition,

$$\bar{\sigma} = (\sigma_1 - \sigma_3)$$

$$(23)$$

and

$$\bar{\varepsilon}^c = \frac{2}{3}(\dot{\varepsilon}^c_1 - \dot{\varepsilon}^c_3)$$

$$(24)$$

where σ_1 and σ_3 are the maximum and minimum principal stresses, respectively. The Von-Mises definition can be regarded as a root mean square maximum shear stress criterion and the Tresca definition as a maximum shear stress criterion. Most investigations of equivalent stress criteria were carried out mainly on thin walled cylinders subjected to different combinations of tension, torsion and internal pressure [21]. The case of internal pressure alone will now be considered as an application of the complex stress creep analysis.

It can be shown for any complex stress state that the equivalent stress calculated according to the Tresca definition is always greater than, or equal to, that determined from the Von-Mises definition. The maximum ratio between them is the $2/\sqrt{3}$ obtained in this example. Use of the Tresca criterion will, therefore, always produce the same, or a higher, creep rate than is obtained from the Von-Mises criterion. As most experimental results usually fall between the two predictions [20-26] assumption of the Tresca criterion is therefore likely to be conservative.

1.5 Damage mechanics concepts

The descriptions presented so far have been concerned mainly with determining the rate of creep deformation in the absence of damage accumulation. Modes of fracture have been identified but the mechanisms by which failure occurs have not been discussed. Two largely consistent damage accumulation processes exist [24,28,30]. They show how tertiary creep is produced in materials that deform in secondary creep in the absence of damage. Both approaches assume that secondary creep can be described by the power law stress dependence given by equation (1).

A model for the development of damage by void growth has been proposed by Cocks and Ashby [28] and extended by Smith and Webster [29]. With this approach a distribution of circular cross-section cavities is assumed on grain boundaries which are perpendicular to the maximum tensile stress. The fractional area of grain boundary damaged therefore is calculated and upper bounds to the displacement rate of a volume of material associated with it can be calculated.

With a phenomenological approach the damage fraction ω is not a physically identifiable quantity. It is an effective fractional loss in area which can be defined to include loss of creep strength due to microstructural degradation as well as cavity nucleation and growth. Empirical equations have been proposed by Kachanov [30] and Rabotnov [24] which allow strain rate and damage rate to increase with this damage ω according to

$$\dot{\varepsilon}^c = \dot{\varepsilon}_0 \left(\frac{\sigma}{\sigma_0} \right)^n \cdot \frac{1}{(1-\omega)^m} = \frac{\dot{\varepsilon}_s^c}{(1-\omega)^m}$$

and

$$\dot{\omega} = \dot{\omega}_0 \left(\frac{\sigma}{\sigma_0} \right)^v \cdot \frac{i}{(1-\omega)^\eta}$$

(25)

where $\dot{\varepsilon}_0$, σ_0, $\dot{\omega}_0$, n, m, v and η are material constants that are chosen to give a best fit to experimental creep data. However, for simplicity it is usually assumed that $m=n$ and $\eta=v$. and further substitution and integration produces

$$\varepsilon^c = \varepsilon_f \left[1 - \left(1 - \frac{t}{t_r} \right)^{1/\phi} \right]$$

(26)

where $\phi = (v + 1)/(v + 1 - n)$. However, for this equation to be physically realistic it is necessary that $v + 1 > n$. In this circumstance and rupture life

$$t_r = \frac{1}{\omega_0} \left(\frac{\sigma_0}{\sigma} \right)^v \frac{1}{(1+1)}$$

(27)

and the creep failure strain is

$$\varepsilon_f = \dot{\varepsilon}_s^c t_r \phi = \varepsilon_{f0} \left(\sigma / \sigma_0 \right)^{n-v}$$

(28)

where ε_{f0} is the creep ductility at stress $\sigma = \sigma_0$.

1.6 Influence of state of stress

In many cases uni-axial analysis is not sufficient for life assessment analyses. However this can be extended to multi-axial loading by accounting for the influence of state of stress on the deformation and damage processes. Several specific models have been proposed which have been discussed by Riedel [10]. They generally assume that strain rate is governed by an equivalent stress criterion and that void growth mechanisms are sensitive to the maximum principal stress or hydrostatic stress component σ_m. For example, Rice and Tracey [31] have developed an expression for rigid-plastic deformation ($n=\infty$), which gives the creep ductility under complex stress loading, ε_f^*, in terms of the uniaxial ductility as

$$\frac{\varepsilon_f^*}{\varepsilon_f} = 1.65 \, exp\left(\frac{-3\sigma_m}{2\overline{\sigma}}\right) \tag{29}$$

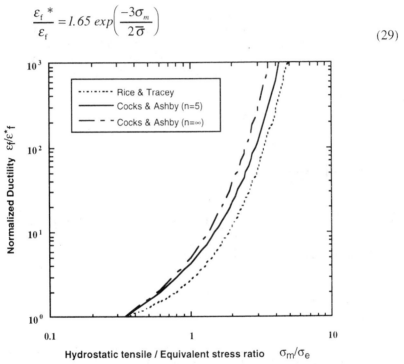

Figure 7: Influence of state of stress on creep failure.

Alternatively, Cocks and Ashby [28] used a void growth model to obtain

$$\frac{\varepsilon_f}{\varepsilon_f^*} = sinh\left[2\left(\frac{n-\frac{1}{2}}{n+\frac{1}{2}}\right)\frac{\sigma_m}{\overline{\sigma}}\right] / sinh\left[\frac{2}{3}\left(\frac{n-\frac{1}{2}}{n+\frac{1}{2}}\right)\right] \tag{30}$$

These ratios are shown plotted in Figure 7. It is apparent that an increase in hydrostatic tension causes a significant reduction in creep ductility for both expressions. This is because an increase in hydrostatic tension, at a constant equivalent stress, enhances the void growth rate without changing the deformation rate causing failure to occur at a lower strain. The significance of this observation on crack growth predictions will be considered later.

1.7 Influence of fatigue

Until now failure by creep processes alone has been examined. Even when introducing the concept of a mechanical equation of state for dealing with variable loading conditions, it was assumed that any influence of fatigue was negligible. A fatigue failure occurs by the initiation and growth of a crack caused by the repeated application of a stress that is less than the tensile strength of the material [32-35]. The fatigue properties of materials are measured mainly in push-pull or reverse bend tests. In general the number of cycles to failure N is sensitive to the stress amplitude σ_a, mean stress σ_m, frequency, wave shape and temperature.

Typically, in a defect free component, most of the life is spent in initiating the crack. Crack initiation occurs by the motion of dislocations up and down slip planes, which gradually results in the formation of extrusions and intrusions on the surface. In the absence of corrosion or environmental attack, fatigue fractures at room temperature are transgranular and stage I occupies the longer fraction of life. At elevated temperatures, intergranular cracking can be observed [32]. When this is seen, the stage I region tends to be reduced and can be eliminated altogether if a grain boundary acts as a crack starter.

The fatigue properties of a material are dependent on its tensile strength and stress-strain characteristics. Alloying additions, heat-treatments and fabrication processes that impede dislocation motion will inhibit the formation of extrusions and intrusions and will tend to improve fatigue lifetimes. Conversely, surface scratches and stress concentrations will tend to reduce lifetimes.

Cumulative damage laws are available for dealing with superimposed mean and cyclic loading and for assessing the influence of variable amplitude cycling. Most engineering components experience variable amplitude loading during operation. It is usually supposed that fatigue damage incurred under these conditions can be accommodated using Miner's cumulative damage law. This states that failure occurs when the fractional damage accumulated at each condition sums to one,

i.e. $$\frac{n_1}{N_1} + \frac{i_2}{i_2} + \frac{n_3}{N_3} + \ldots = 1$$

or $$\sum \frac{i}{N} = 1$$

(31)

where n_1, n_2, n_3 etc. are the number of cycles spent at each condition and N_1, N_2, N_3 etc. are the endurance at these conditions. The relation should be used as a general guide only as experimental evidence indicates that the fractional damage suffered at failure can range between about 0.5 and 2.0. This can usually be attributed to sequence of loading effects which are not accommodated in a simple expression like Miner's Law.

Under combined static and cyclic loading at elevated temperatures creep and fatigue processes can take place together [36-37]. It is possible for these processes to occur independently or in conjunction depending upon the controlling mechanisms in each case.

In the former case failure will be dominated by the process which occurs most rapidly and in the latter by the slowest process. When creep mechanisms govern it is likely that the fracture surface will be intergranular and when fatigue processes dominate transgranular. When both mechanisms contribute to failure a mixed mode of fracture may be expected.

It has been proposed that when creep and fatigue processes occur independently, that the fractions of damage incurred by each mechanism can be summed separately using equation (19) and (31) to predict failure when,

$$\sum \frac{i}{t_r} + \sum \frac{n}{N} = 1$$

(32)

In this equation t_r is the anticipated rupture life due to creep alone. Other approaches have been proposed for dealing with more complicated interaction effects. The two most common are due to Coffin [37] and Manson [38]. In the former a frequency term is incorporated to allow for time dependent effects. In the latter a strain range partitioning (SRP) approach is adopted for separating the total strain range each cycle into creep, fatigue and mixed creep and fatigue components. With this method, a total of four components can be identified depending upon the type of loading cycle and whether hold times are present.

2. Concepts of Fracture Mechanics

The previous section dealt with creep of uncracked bodies. Sections 3 and 4 will consider the cracked body and its analysis using fracture mechanics concepts. Figure 8 shows schematically the various regimes of fracture found both at room temperature and at elevated

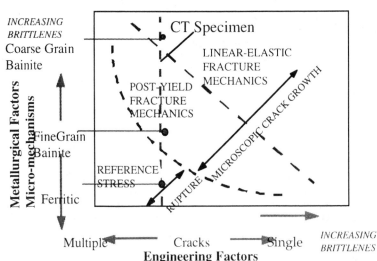

Figure 8: Schematic view of the regimes of fracture in terms of geometric constraint and metallurgical factors ranging from elastic to fully ductile behaviour.

Temperature. Generally the sharper and the larger the grains the more brittle the material behaviour and the more likely that Linear Elastic Fracture Mechanics (LEFM) will be the relevant correlating parameter. Conversely the more ductile the material and the more diffuse the number of cracks or damage region it is likely that Non-Linear LEFM will describe the crack tip region. The assumptions hold for creep with the difference only that in creep deformation occurs in time and in plasticity it is only load dependent. The crack tip stress singularity decreases with increasing ductility and in the extreme the stress at the crack tip is equivalent to the remote stress and the analysis would use net-section stress or reference stress σ_{ref} to predict failure life.

The first analytical study of fracture mechanics was made in the 1920s [12] for cracks in an ideal brittle material. This study was based on the consideration of energy balance in a cracked body. The stress intensity factor K was proposed, as a parameter that describes the stress state around a crack tip, by Irwin [39]. The energy release rate G which corresponds uniquely to stress intensity factor K was also proposed. These fracture mechanics parameters in LEFM have since been used n engineering procedures. In ductile materials with low yield stress or in creeping materials the LEFM approach is not sufficient in predicting fracture of a cracked structure made of ductile or creeping materials. The J contour integral [40-41] was proposed as a fracture mechanics parameter for estimating ductile fracture. The J integral has been widely accepted as an effective parameter for the assessment of ductile fracture for large scale yielding cases.

Fracture induced by a crack is controlled by the stress and strain state in the failure process zone in the vicinity of a crack tip. The fracture mechanics approach characterises the crack tip stress and strain fields by the stress intensity factor K or the J integral. The stress and strain fields characterised by these fracture mechanics parameters are asymptotic fields at a crack tip and are not exact solutions around the crack tip. For creep a parameter called C* is developed which is analogous to J but is dependent on strain rate rather than strain due to loading. Figure 9 gives a description where the various parameters would be applicable.

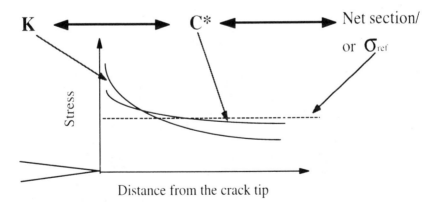

Figure 9: Schematic profile of the stress field at the crack tip and the relevant fracture mechanics parameters associated with them.

2.1 Linear elastic fracture mechanics

Stress intensity factor K is a parameter for characterising the local stress and strain fields around a crack tip based on linear elastic analysis. The stress state near a crack tip is expressed using stress intensity factors by;

$$\sigma_{(r,\theta)} = \frac{K_I}{\sqrt{2\pi r}} f^I(\theta) + \frac{K_{II}}{\sqrt{2\pi r}} f^{II}(\theta) + \frac{K_{III}}{\sqrt{2\pi r}} f^{III}(\theta)$$

(33)

where r, θ are the polar co-ordinates ahead of a crack tip as shown in Figure 8, and K_I, K_{II} and K_{III} are stress intensity factors for mode I, mode II, and mode III loading. Mode I is the opening mode, mode II is the in-plane shear mode, and mode III is the transverse shear mode. The function $f(\theta)$ is a function of angle θ for each mode.

It is usual for mode I loading to be used in analyses since this is the most common loading mode which causes crack extension at elevated temperature. The mode I stress fields around a crack tip analysed elastically are expressed by the following equations [42].

$$\sigma_x = \frac{K_I}{\sqrt{2\pi r}} \cos\frac{\theta}{2}\left(1 - \sin\frac{\theta}{2}\sin\frac{3\theta}{2}\right) + \text{high order terms}$$

(34)

$$\sigma_y = \frac{K_I}{\sqrt{2\pi r}} \cos\frac{\theta}{2}\left(1 + \sin\frac{\theta}{2}\sin\frac{3\theta}{2}\right) + \text{high order terms}$$

(35)

$$\tau_{xy} = \frac{K_I}{\sqrt{2\pi r}} \cos\frac{\theta}{2}\sin\frac{\theta}{2}\cos\frac{3\theta}{2} + \text{high order terms}$$

(36)

$$\sigma_z = v\left(\sigma_x + \sigma_y\right) \quad \text{for plane strain}$$

$$= 0 \qquad\qquad \text{for plane stress}$$

where each stress component is shown in Figure 10.

The high order terms are usually ignored in characterising the stress state local to a crack tip because they do not have a singularity near the crack tip, while the first term shows a $r^{-1/2}$ singularity. When only the first term is used, the stress approaches zero as the distance becomes large, while it should go to σ, the ligament stress. This means that the stress field characterised by K neglecting the high order term is valid only for the area around the crack tip, where the singular term dominates the stress field. As r becomes smaller, the stress becomes infinite according to this characterisation, but this is not representative of the real behaviour because plastic deformation around a crack tip keeps the stress finite. Nevertheless, if the stress fields outside the plastic zone are the same in any structures of the same material, the material behaviour within the plastic zone should be the same. Therefore, the stress intensity factor can still be an adequate parameter provided that the plastic zone is small enough compared to the K dominant stress field area. This is called the small scale yielding condition.

The stress intensity factor K should be expressed by the following form because of dimensional considerations in equation (33).

$$K = Y\sigma\sqrt{a}$$
$$\text{or } K = F\sigma\sqrt{\pi a}$$

(37)

where Y and F are non dimensional factors, σ is nominal stress and a is crack length. For a crack within an infinite plate, the Y factor is given as $\sqrt{\pi}$ (or $F = 1$), then K value becomes;

$$K = \sigma\sqrt{\pi a}$$

(38)

The Y factor is generally dependent on crack sizes and component dimensions, as well as loading and boundary conditions. The Y values have been calculated analytically or numerically for various geometry and loading conditions, and they are available in reference handbooks, for example [44-46].

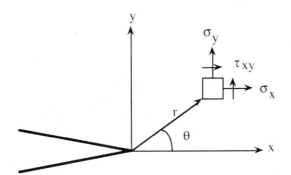

Figure 10. Co-ordinates and stresses around a crack tip

2.2 Energy release rate interpretation

The energy release rate concept is based on the Griffith energy criterion of fracture [47]. The idea is that "crack growth can occur if the energy required to form an additional crack of size da can just be delivered by the system energy balance." Then the condition of crack growth becomes,

$$\frac{dU}{da} = \frac{d}{da}(U_c - U_s) = \frac{dW}{da}$$

(39)

where U is the potential energy of a cracked body, U_c is the work performed by external forces, U_s is the internal energy stored in the body as strain energy, and W is the energy required for forming a new crack surface. The potential energy change due to crack extension dU/da per unit thickness is called the energy release rate and is designated by G.

$$G = \frac{dU}{B\,da}$$

(40)

where B is the thickness of the cracked body. The energy release rate concept is illustrated in Figure 11 for crack extension at constant load. The relationship between the energy release rate and stress intensity factor is given by the following equations for mode I loading.

$$G = \frac{K_I^2}{E'}$$

(41)

where $E' = E$ for plane stress $E' = E/(1-v^2)$.   for plane strain   (42)

and E is the Young's modulus and v is Poisson's ratio. The energy release rate G can be used as a fracture mechanics scalar value if the cracked body generally behaves linearly elastic.

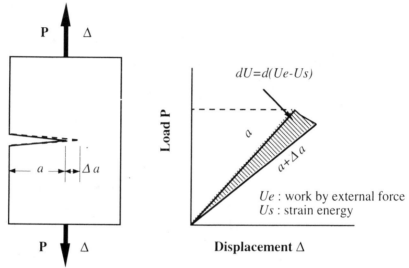

Figure 11. Energy release rate concept

2.3 Non-linear fracture mechanics

The linear elastic fracture mechanics concept is an appropriate parameter for relatively brittle materials used in large structures. If the global deformation of a cracked structure is not linear due to plasticity or creep, LEFM will not be able to describe adequately the stress state close to a crack tip, or energy conservation in a cracked body. In these circumstances it will be necessary to employ non-linear fracture mechanics (NLFM).

2.4 The J integral concept

The J integral can be interpreted in several ways. From the view of fracture energy, the J integral can be considered as an extension of the linear elastic energy release rate G for elasto-plastic material. J equals G in linear elastic cases. Precisely speaking, the J integral is defined for non-linear elasticity (often called deformation plasticity), not for general plasticity in which plastic strain is influenced by the loading history. Nevertheless, the J integral is used for most practical plastic materials and is applicable when unloading does not take place. In the following sections, the plastic and non-linear elastic cases are taken to be identical unless otherwise stated.

As an energy parameter, the J integral is the energy required to form new crack surfaces which is equal to the potential energy change due to crack extension. A definition of the integral can be made in the same way as for the energy release rate G by;

$$J = \frac{1}{B}\frac{dU}{da} = \frac{1}{B}\frac{d}{da}(U_e - U_s)$$

(43)

where the U, U_e and U_s are the same energies defined above Figure 12 shows the potential energy change dU for crack growth da.

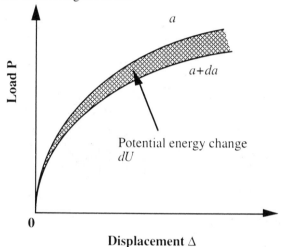

Figure 12. J integral concept form the view of energy

The change in the area dU is the potential energy change needed for the crack to extend by da. It should be noted that in this case the available potential energy does not fully transfer to crack extension as the energy will also be lost due to plastic deformation.

The J integral can be also defined as a particular contour integral that is equal to the energy release rate in non-linear elastic materials as;

$$J = \int_\Gamma \left(W dy - T_i \frac{\partial u_i}{\partial x} ds \right)$$

(44)

where Γ: Line integration path as shown in Figure 13.s: is the Length along the path Γ, W the Strain energy density (non-linear elastic strain energy as shown in Figure 14,

$$W = \int_0^{\varepsilon_{ij}} \sigma_{ij} d\varepsilon_{ij}$$

(45)

T_i: Traction force

$$T_i = \sigma_{ij} n_j$$

(46)

n_j: Normal unit vector outside from the path Γ (See Figure 14.) This contour integral is a conservation integral and provides the path independent values when contours are taken around a crack tip [41]. As the elastic energy release rate G is related to K, the J integral, which is also energy release rate in non-linear materials, can also characterise the stress and strain fields around a crack tip. When non-linear behaviour of a material is expressed by a power law as,

$$\left(\frac{\varepsilon}{\varepsilon_0}\right) = \left(\frac{\sigma}{\sigma_0}\right)^n \tag{47}$$

where σ_0, ε_0 and n are material constants, the stress and strain at a distance r from crack tip

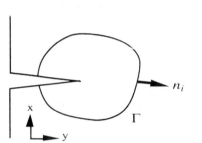

Figure 13: Integration Path for J integral   Figure 14: Strain Energy W

are predicted by the following equations using the J integral. The stress and strain fields are often called the HRR fields. [40-46]

$$\sigma_{ij} = \sigma_o \left(\frac{J}{I_n \sigma_0 \varepsilon_0 r}\right)^{1/(n+1)} \tilde{\sigma}_{ij} \tag{48}$$

$$\varepsilon_{ij} = \varepsilon_o \left(\frac{J}{I_n \sigma_0 \varepsilon_0 r}\right)^{n/(n+1)} \tilde{\varepsilon}_{ij} \tag{49}$$

where I_n is the non-dimensional factor of n, and σ_{ij} and ε_{ij} are functions of angle θ and n, and are normalised to make their maximum equivalent stress and strain unity. Figure 9 shows schematically how the stresses would vary with increasing N. N=1 is the linear elastic case and the larger n values reduce the singularity accordingly. Calculated value of I_n. [40,46,47] show that it varies between 2-6 for a range of n between plane stress to plane strain [46,47]. McClintock [47] gave approximated equations for I_n values as,

$$I_n = 10.3\left(0.13 + \frac{1}{n}\right)^{1/2} - \frac{4.8}{n} \text{ for plane strain} \tag{50}$$

Because the stress and strain fields given by equations (48) and (49) are based on small-deformation theory, the true state will be different close to a crack tip where the assumption of small-deformation theory is not adequate. For this reason, the stress or strain field characterised using the J integral is not the actual state at a crack tip itself. However, the J integral can still be considered as a valid fracture parameter as long as the fracture process zone is related to the surrounding J integral stress and strain fields. This is similar to the concept of applying the stress intensity factor for describing fracture in a material which undergoes small scale yielding.

2.5 Procedures to calculate J

Procedures to calculate J integrals are outlined in this section. At first, numerical procedures are introduced as well as simplified methods for engineering use. The procedures to obtain the J integral from experimental results are also mentioned briefly.

The numerical calculation of the J integral is mainly categorised into the following two methods. Contour integration and virtual crack extension. The first method follows the numerical definition in equation (44). The J integral can be calculated from the stress, strain and displacement along an arbitrary path Γ. This method is often employed in post processors for finite element calculations. The applicability in three dimensional cases is not clear because the definition of equation (44) was originally based on two dimensional considerations.

The second procedure as incorporated into commercial finite element codes such as ABAQUS is the virtual crack extension (VCE) method which is easier to expand to three dimensional cases. The fundamental concept of the VCE method is based on the energy release rate concept formulated in equation (43) [49,50].

2.6 Numerical method to calculate J

Although the numerical methods to analyse the J integral are becoming easier to apply, owing to improvements in computer performance and software, including pre/post processing systems, simplified methods are still desired for engineering use of NLFM. Because the problem is non linear, it is generally difficult to obtain the J integral by analytical methods. Analytical solutions for only very simple cases are available, for example a 2-D crack in an infinite plate with power law plasticity[51]. Therefore, solutions for J integrals for various geometries and material properties have to rely on numerical calculations. The simplified methods are usually, themselves, based on previous numerical solutions. The most common engineering approach for calculating the J integral which is based on the GE / EPRI method [52]. In this method, the elasto-plastic material property is expressed by a Ramberg-Osgood type of formula as;

$$E\varepsilon = \sigma + \alpha \left(\frac{|\sigma|}{\sigma_0} \right)^{n-1} \sigma \qquad (51)$$

where E is Young's modulus, α, σ_0 and n are material constants. The first term expresses the elastic component, and the second term is the plastic component. As the simplest combination of an elastic part and a plastic part, the J integral is calculated by summation of the elastic J_{el} and the plastic J_{pl} integrals.

$$J = J_{el} + J_{pl} \tag{52}$$

The J_{el} is the same as the energy release rate G and can be calculated from the stress intensity factor K using equation (41). The plastic part of J_{pl} is given by;

$$J_{pl} = \alpha \, \sigma_0 \, \varepsilon_0 \, c \, h_1 \left(\frac{P}{P_0} \right)^{n+1} \tag{53}$$

where $\varepsilon_0 = \sigma_0/E$ and c is characteristic length, usually taken as the uncracked ligament length. The P and P_0 are an applied load and characterising load, respectively. h_1 is a non dimensional function of geometry and n and has been tabulated for specific values of P_0 [52-58]. Any P_0 value can be used, but to give relatively constant h_1 values against change of crack depth a and the material constant n, usually the P_0 is taken as plastic collapse load or limit load [52-58].

This method (called the GE/EPRI) is applicable to simple geometries like a fracture mechanics test specimens of pipe components. However, if a fracture mechanics estimate for an other geometry is required, numerical calculations are required using fully plastic material properties to make new h_1 tables. Some further calculations have been made to extend the h_1 tables, especially for 3 dimensional surface cracks that are realistic defects in actual components, for example [59].

2.7 Reference stress method

Another engineering approach to calculate the J integral has been proposed using reference stress procedures [60-62] the reference stress has been also utilised for estimating non-linear fracture mechanics parameters. The reference stress is defined by,

$$\sigma_{ref} = \sigma_y \frac{P}{P_{LC}} \tag{54}$$

where P is the applied load, and P_{LC} is the plastic collapse load of the cracked body made of elastic perfect plastic material with yield stress σ_y. This reference stress concept has been extended to be used in non-linear fracture mechanics estimations [62]. The basic idea of the reference stress method was originally modification of the GE/EPRI technique mentioned in the previous section. A stress σ_r is defined by

$$\sigma_r = \sigma_0 \left(\frac{P}{P_0} \right) \tag{55}$$

to produce a plastic strain component ε_r from equation (51).

$$\varepsilon_r = \alpha \, \varepsilon_0 \left(\frac{P}{P_0} \right)^n \tag{56}$$

Substituting equations (54) and (55), equation (53) becomes;

$$J = \sigma_r \, \varepsilon_r \, c \, h_1 \tag{57}$$

If an appropriate P_0^* is chosen instead of P_0 for equation (53), the h_1 values become approximately independent of the material constant n.

$$h_{1,n} = h_{1,1} \tag{58}$$

where the second subscripts n and 1 show the values of the power law exponent. For elastic $n = 1$ condition, the J integral is related to stress intensity factor K by equation (41), so that equations (56) and (57) give;

$$\frac{K^2}{E'} = \sigma_r \, \varepsilon_r \, c \, h_{1,1} \tag{59}$$

where E' is defined by equation (42). Since ε_r is also related to σ_r for the elastic case by;

$$\varepsilon_r = \sigma_r / E \tag{60}$$

the following relationship is obtained.

$$c \, h_{1,n} = c \, h_{1,1} = \frac{K^2}{E' \sigma_r \varepsilon_r}$$

$$= \frac{K^2}{\sigma_r^2} \qquad \text{for plane stress}$$

$$= \frac{(1 - v^2) K^2}{\sigma_r^2} \qquad \text{for plane strain} \tag{61}$$

Using this relation in equation (57) gives;

$$J = \mu \, \sigma_r \, \varepsilon_r \left(\frac{K}{\sigma_r} \right)^2 \tag{62}$$

where $\mu = 1$ for plane stress

$$\mu = (1 - v^2) = 0.75 \text{ for plane strain} \tag{63}$$

To obtain the relations above, an appropriate P_0 value must be chosen. It has been concluded that the plastic collapse load P_{LC} is a good approximation of P_0^* for examining the dependence of h_1 values on n for several geometries. Substituting P_{LC} for P_0, the stress σ_r becomes σ_{ref}, then the J integral is calculated using the reference stress as;

$$J = \mu \sigma_{ref} \varepsilon_{ref} \left(\frac{K}{\sigma_{ref}} \right)^2 \tag{64}$$

where ε_{ref} is the uniaxial strain corresponding to the reference stress σ_{ref}. It should be noted that, as the reference stress estimation is an approximate solution, the effect of μ is relatively small compared to other factors and often ignored.

There are advantages of the reference stress method over the original GE/EPRI method. Power law fitting of actual plastic behaviour is not necessary since is not easy to fit a power law equation to an actual stress and strain curve on wide range of plastic behaviour. The GE/EPRI equation (53) has a strong dependency on the plastic exponent n and hence the J integral. The reference stress method can use actual stress and strain relationships. Also the reference stress method does not require the h_1 values that are specific to individual geometries and material properties and need to be derived using numerical methods whereas the reference stress method only needs plastic collapse loads of the specific geometries. Collapse loads solutions are available for a number of geometries and loading in the literature, for example [63-64]. Consequently, the reference stress method is easier to employ for flaw assessments in general structures.

2.8 Experimental J integral

The J integral in fracture mechanics experiments can be estimated directly from experimental measurements. Procedures to estimate experimental J integrals are available in an ASTM standard [65]. In this section, the procedure for standard compact tension specimens is briefly introduced. Calculations of the J integral are made from load and the load point displacement curve. The area under the load-displacement curve is converted to energy units according to the load scale and displacement scale. The J integral is calculated as a summation of the elastic component J_{el} and the plastic component J_{pl}.

$$J = J_{el} + J_{pl} \tag{65}$$

The elastic component is calculated from the stress intensity factor K by;

$$J_{el} = \frac{K^2(1-v^2)}{E} \tag{66}$$

where
$$K = \frac{P}{B_e\sqrt{W}} \cdot f(a/W) \tag{67}$$

$$B_e = \sqrt{B \cdot B_n} \tag{68}$$

$$f(a/W) = \frac{(2+a/W)\{0.886 + 4.64\,a/W\,13.32(a/W)^2 + 14.72(a/W)^3 - 5.6(a/W)^4\}}{(1-a/W)^{3/2}} \tag{69}$$

for a specimen of width W and thickness B containing side grooves to give a thickness between side grooves of B_n. The plastic component is obtained by;

$$J_{pl} = \frac{\eta A}{B_n(W-a)} \tag{70}$$

where $\eta = 2 + 0.522(1 - a/W)$ \hfill (71)

For other specimen geometries, the same concept can be applied. The stress intensity factor K and geometry factors η will be modified for each specimen geometry. Equation (70) is an approximate estimate of the J integral using an analytically obtained function η. This method has an advantage that J can be estimated from only one experimental P - Δ curve.

3. Concepts of High Temperature Fracture Mechanics

The arguments for high temperature fracture mechanics essentially follow those presented in the fracture mechanics sections 2 and 3. For creeping situations where elasticity dominates the stress intensity factor may be sufficient to predict crack growth. However as creep is a non-linear time dependent mechanism even in situation where small scale creep may exist linear elasticity my not be the answer. By using of the J definition estimation procedures are used in the developing the fracture mechanics parameter $C*$. Methods are then shown for its evaluation.

A simplified expression for stress dependence of creep is given by a power law equation which is often called Norton's creep law;

$$\dot{\varepsilon} = C\sigma^n$$

$$\text{or} \quad \frac{\dot{\varepsilon}}{\dot{\varepsilon}_0} = \left(\frac{\sigma}{\sigma_0} \right)^n \tag{72}$$

where C, n, ε_0 and σ_0 are material constants. This equation is used to characterise the steady state (secondary) creep stage where the hardening by dislocation interaction is balanced by recovery processes. The typical value for n is between 3 and 10 for most metals. The C value is a function of temperature.

3.1 The C* integral concept

The stress fields characterised by K in elasticity will be modified to the stress field characterised by J in plasticity in the region around the crack tip. In the case of large scale creep where stress and strain rate determine the crack tip field other fracture mechanics parameters are needed and in this case the parameter $C*$ which is analogous to J has been proposed [66] for this purpose. The $C*$ integral has been widely accepted as the fracture mechanics parameter for this purpose. $C*$ integral is a parameter of the J integral. The definition of the $C*$ integral is obtained by substituting strain rate and displacement rate for strain and displacement of the J integral defined by equation (44) as,

$$C* = \int_{\Gamma} \left(\dot{W}dy - T_i \frac{\partial \dot{u}_i}{\partial x} ds \right) \tag{73}$$

where $\dot{W}$ is strain energy density change rate,

$$\dot{W} = \int_0^{\varepsilon_{ij}} \sigma_{ij}d\dot{\varepsilon}_{ij}$$

$$\dot{\varepsilon}_{ij} = \frac{d\varepsilon_{ij}}{dt} \tag{74}$$

and $\dot{u}_i$ is displacement rate ($= du_i / dt$). The other notations are the same as in the J integral definition. As the J integral characterises the stress and strain state, the $C*$ integral is also expected to characterise the stress and strain rate around a crack. For a non-linear elastic material, the asymptotic stress and strain fields are expressed by equations (48) and (49). Due to the analogy between non-linear elasticity and non-linear viscosity, the stress and strain rate in materials following Norton's creep law by equation (72) are given as;

$$\sigma_{ij} = \sigma_o \left(\frac{C^*}{I_n \sigma_0 \dot{\varepsilon}_0 r} \right)^{1/(n+1)} \tilde{\sigma}_{ij} \tag{75}$$

$$\dot{\varepsilon}_{ij} = \dot{\varepsilon}_o \left(\frac{C^*}{I_n \sigma_0 \dot{\varepsilon}_0 r} \right)^{n/(n+1)} \tilde{\varepsilon}_{ij} \tag{76}$$

Therefore the stress and strain rate fields of non-linear viscous materials are also HRR type fields with $\tilde{\sigma}_{ij}$ and $\tilde{\varepsilon}_{ij}$ defined as before. By analogy with the energy release rate definition of J, the $C*$ integral can be obtained from

$$C^* = \frac{1}{B} \frac{dU^*}{da} \tag{77}$$

where U^* is the rate of change of the potential energy dU/dt shown in Figure 13. From the view of an energy balance, the $C*$ integral is the rate of change of potential energy with crack extension.

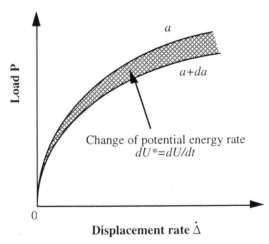

Figure 15. $C*$ integral concept from the view of energy balance

3.2 Experimental estimates of the C* integral

As mentioned in the previous section, the $C*$ integral is an analogous parameter to the J integral for a plastic material. The experimental $C*$ integral can therefore be calculated by substituting displacement rate for displacement in the equation for J [65]. Hence the $C*$ integral can be obtained from;

$$C* = \frac{n}{n+1} f(a/W) \frac{P \dot{\Delta}_c}{B_n W} = F \frac{P \dot{\Delta}_c}{B_n W} \qquad (78)$$

where P is the load, $\dot{\Delta}_c$ is the displacement rate due to creep, B_n is the net thickness between side grooves, and W is the width of the specimen. The $f(a/W)$ is a non-dimensional geometry function of a crack length a. For compact tension specimens, the $f(a/W)$ is given by analogy with fracture toughness J_{IC} testing in ASTM E813 [65] shown in the previous chapter.

$$f(a/W) = \frac{2 + 0.552(1 - a/W)}{1 - a/W} \qquad (79)$$

Equations (78) and (79) are also introduced in a new ASTM standard for the measurement of creep crack growth rates in metals using compact tension specimens [40]. For other specimen geometries, the geometry function $f(a/W)$ is given by several researchers [66,67]. An approximate estimation of the $f(a/W)$ can be made using plastic collapse loads.[68]. The change of applied load due to crack extension at constant displacement is assumed to be proportional to the change of the collapse load, and the resulting equation for the function $f(a/W)$ is written as;

$$f(a/W) \approx -\frac{W}{P_{LC}} \cdot \frac{dP_{LC}}{da} \qquad (80)$$

where a and P_{LC} are crack length and collapse load of the specimen respectively. This approach is very convenient for engineering use because plastic collapse loads are given for a wide range of geometries in the literature, for example [64].

3.3 Numerical method

Procedures for the numerical calculation of the $C*$ integral are the same as those for the J integral mentioned earlier except that strain rate and displacement rate replace strain and displacement, respectively. The $C*$ integral can be calculated by a contour integration method.

From the consideration of energy balance in equation (77), the $C*$ integral can also be obtained from change of J integral values. The $C*$ integrals are derived from change in the J integral using the following relationship between $C*$ and J.

$$C^* = \frac{1}{B}\frac{dU^*}{da} = \frac{1}{B}\frac{d}{da}\left(\frac{dU}{dt}\right) = \frac{d}{dt}\left(\frac{1}{B}\frac{dU}{da}\right) = \frac{dJ}{dt} \tag{81}$$

When the power law creep strain equation (Norton's creep law) is employed, the C^* integral can be obtained directly from J. If the exponent n for power law plasticity shown in equation (47) is the same as the creep exponent for Norton's law, the stress state of a fully plastic body and steady state creeping body will be the same. Hence the strain distribution in a plastic body is proportional to the strain rate distribution of a creeping body. When plastic strain ε_p is expressed by the power law,

$$\varepsilon_p = A\sigma^n \tag{82}$$

and Norton's creep law by equation (72), the ratio of strain rate or displacement rate for a steady creep condition, and strain or displacement in a fully plastic analysis is C/A. Hence by substitution, the relationship between C^* and J is obtained as follows.

$$C^* = \frac{C}{A}J \tag{83}$$

Equation (83) suggests that only a J calculation for fully plastic behaviour is required to calculate the C^* integral for steady state creep conditions.

3.4 GE/EPRI method

In the case of a power law material, there exists the simple relationship in equation (83) between J and steady state C^* as mentioned above. Therefore, the GE/EPRI method introduced earlier in the engineering approach for calculating the J integral, is also applicable for calculating C^*. When power law plasticity is used, the plastic J integral can be expressed using a non-dimensional factor h_1 and material constants as in equation (49). If the creep exponent is the same as the plastic exponent n, the C^* integral can be obtained using strain rate instead of strain as follows;

$$C^* = \sigma_0 \dot{\varepsilon}_0 c h_1 \left(\frac{P}{P_0}\right)^{n+1} \tag{84}$$

where h_1 is common between the C^* and J integrals. This method is very useful for engineering applications of the C^* integral because it does not require a non-linear stress and strain analysis. However the same limitation exists as for the J integral, and it is necessary to have the non dimensional factor h_1 for a specific geometry and material constant n.

Reference stress method

A reference stress method for C^* calculation is also easily obtained by analogy with that for the J integral introduced earlier. The only difference is the strain ε_{ref} is replaced by the strain rate $\dot{\varepsilon}_{ref}$ at a reference stress σ_{ref}. The C^* formulation using reference stress becomes

$$C^* = \mu \sigma_{ref} \dot{\varepsilon}_{ref} \left(\frac{K}{\sigma_{ref}} \right)^2 \tag{85}$$

The μ and K have the same definitions as previously.

3.5 Transition time and the C(t) integral

In the previous sections, the C^* integral was calculated analogously from the J integral. Since C^* is a parameter that characterises the stress and strain rate field in a cracked body where full stress redistribution has been completed, stress and strain state may be different from the state predicted using C^* during transient conditions. Hence, the C^* integral would not be an accurate parameter to predict the behaviour around a crack tip while stress is still redistributing. For instance, considering the sudden application of a constant load to an elastic non-linear viscous body, the stress at $t = 0$ is purely elastic and the stress field around the crack tip is given by the inverse square root singularity (K stress field). At any $t > 0$ after the sudden loading, the creep strain near the crack tip increases until it becomes much larger than the elastic strain when $n > 1$. Therefore, the stress and strain rate fields will be characterised as an HRR type field similar to that for steady state creep conditions. The only difference from the steady state creep is that the characterising parameter C^* is replaced by a transient parameter $C(t)$ when the creep dominant region is surrounded by an elastic stress field, and stress and strain rate fields are given by, [69]

$$\sigma_{ij} = \sigma_o \left(\frac{C(t)}{I_n \sigma_0 \dot{\varepsilon}_0 r} \right)^{1/(n+1)} \tilde{\sigma}_{ij} \tag{86}$$

$$\dot{\varepsilon}_{ij} = \dot{\varepsilon}_o \left(\frac{C(t)}{I_n \sigma_0 \dot{\varepsilon}_0 r} \right)^{n/(n+1)} \tilde{\varepsilon}_{ij} \tag{87}$$

An approximate expression of $C(t)$ is determined by the following equation [70].

$$C(t) = \frac{G}{(n+1)t} \tag{88}$$

where the G is the energy release rate which can be replaced by J in plastic-viscous materials. A transition time t_T to reach steady state has been defined as the time when the $C(t)$ integral equals C^* [43]. Substituting C^* for $C(t)$ in equation (88),

$$t_T = \frac{G}{(n+1)\, C^*} \tag{89}$$

Comparison of the transition time with actual expected duration provides an idea of whether creep crack growth is controlled by K or by C^*. From finite element results, an alternative expression for $C(t)$ using the transition time t_T is approximately estimated as, [71]

$$C(t) = \left(1 + \frac{t_T}{t}\right) C^*$$ (90)

Ainsworth and Budden [72] also gave the modified approximate equation for $C(t)$.

3.6 C_t parameter

Another fracture mechanics parameter C_t for describing the transient creep stage has been proposed by Saxena [73]. While $C(t)$ in the previous section is a parameter to characterise the stress and strain rate fields around a crack tip for the transient creep condition, the C_t parameter is based on energy concepts. For the steady state condition, C^* has both its characterising role (equations (75) and (76)) and its energy interpretation (equation (77)). However, during transient creep C^* is replaced by $C(t)$ in equations (75) and (76) and by C_t in the energy expression,

$$C_t = \frac{1}{B} \frac{dU^*}{da}$$ (91)

Bassani [74] has given an approximate expression for C_t by analogy with equation (90) for $C(t)$ as;

$$C_t = \left[1 + \left(\frac{t_T}{t}\right)^{\frac{n-3}{n-1}}\right] C^*$$ (92)

Creep crack growth data of a 1Cr1Mo$^{1}/_{4}$V steel had been analysed using both C^* and C_t, and a better correlation of the crack growth rate was found with C_t among the different specimen geometries of different thickness with and without side-grooves [74].

3.7 Stress redistribution time

The stress and strain rate fields were determined in the previous section from equations (86) and (87) in terms of $C(t)$. In this section, another approach for characterising stress redistribution is proposed. On initial loading, in the absence of plasticity, the stress distribution will be elastic and will gradually redistribute to its steady state creep condition with time as illustrated in Figure 16. At a distance r^* from the crack tip, a stress σ^* can be defined where the stress remains approximately constant and the elastic and creep stresses have the same magnitude. The value of r^* can be calculated from equations (35) and (75) for plane stress conditions as;

$$r^* = \left(\frac{K^2}{2\pi}\right)^{\frac{n+1}{n-1}} \left(\frac{I_n C}{C^*}\right)^{\frac{2}{n-1}}$$ (93)

Equation (93) identifies the region $r < r^*$ in which stress relaxation takes place at the crack tip because of creep. Equation (93) can be expressed in non-dimensional form using the GE/EPRI expression [75]. Another non-dimensional expression is possible using the reference stress method from equation (85) as;

$$\frac{r^*}{W} = \left\{ \left(\frac{1}{2\pi}\right)^{n+1} \left(\frac{I_n}{\mu}\right)^2 \right\}^{\frac{1}{n-1}} \left(\frac{K}{\sigma_{ref}\sqrt{W}}\right)^2 \qquad (94)$$

The first half of the right side is the term including the influence of n, and the last term is a geometrical factor. If a distance r_{ref} is defined as the distance where the elastic stress calculated by $\sigma = K/\sqrt{2\pi r}$ is equal to reference stress, r_{ref} is written as;

$$\frac{r_{ref}}{W} = \left(\frac{1}{2\pi}\right) \left(\frac{K}{\sigma_{ref}\sqrt{W}}\right)^2 \qquad (95)$$

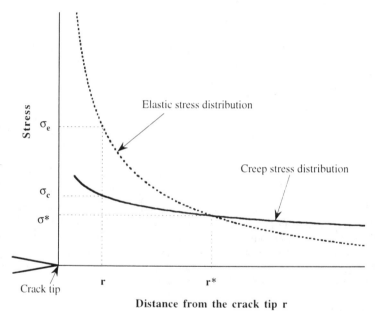

Figure 16: Elastic and steady state creep distribution ahead of the crack tip

Comparing equation (95) with equation (94), r_{ref} is the extreme value of r^* for n =∞. For CT specimens under plane strain conditions, the non-dimensional distance r^*/W is relatively insensitive to n. Substituting typical values of crack length $a = 25$ mm and $W = 50$ mm for tests on standard compact tension specimens gives $r^* \approx 3$ mm. Since the size of the creep process zone r_c in which creep damage accumulates is nominally of the order of a material's grain size [76,77], this implies that creep damage develops well within the region where stress relaxation takes place during the redistribution period. An estimation of the time t_w of this redistribution period can be determined as the time taken for the creep strain ε_c to equal the elastic strain ε_e at $r = r^*$ such that:

$$\varepsilon_e\left(r=r^*, t=t_w\right) = \varepsilon_c\left(r=r^*, t=t_w\right) \qquad (96)$$

Solution of equations (35), (75), (93) and (96) gives the redistribution time t_w as:

$$t_w = \frac{I_n}{2\pi} \frac{K^2}{E'C^*} = \frac{I_n}{2\pi} \frac{G}{C^*} \tag{97}$$

This expression can be compared with the transition time t_T defined by equation (89). The relationship between t_w and t_T is obtained from equations (89) and (97) as;

$$\frac{t_w}{t_T} = \frac{(n+1)I_n}{2\pi} \tag{98}$$

When C^* is estimated using reference stress methods by equation (75), the redistribution time t_w in equation (97) is rewritten as;

$$t_w = \frac{I_n}{2\pi} \frac{\sigma_{ref}}{E\dot{\varepsilon}_{ref}} \tag{99}$$

Another redistribution time t_{red} is used in Nuclear Electric R5 [78] as;

$$t_{red} = \frac{\sigma_{ref}}{E\dot{\varepsilon}_{ref}} \tag{100}$$

Then the ratio of t_w and t_{red} is given from equation (99) and (100) as;

$$\frac{t_w}{t_{red}} = \frac{I_n}{2\pi} \tag{101}$$

Because I_n is relatively insensitive to n, the value of t_w is 80 % of t_{red} for plane strain conditions and 50 % for plane stress conditions for a wide range of n values. The stress at $r = r^*$ is constant during stress redistribution so that the strain rate $\dot{\varepsilon}^*$ at r^*, as well as creep strain rate $\dot{\varepsilon}_c$ at r^*, is also constant for secondary creep because no elastic strain changes occur at this point. If the strain rate at a distance r ahead of a crack tip is expressed as α times the strain rate at distance r^*, the following relation can be obtained.

$$\alpha \dot{\varepsilon}^* = \dot{\varepsilon}_e + \dot{\varepsilon}_c \tag{102}$$

where the $\dot{\varepsilon}_e$ and $\dot{\varepsilon}_c$ are the elastic strain rate and creep strain rate at distance r, respectively. When Norton's creep law is employed, equation (102) can be written as;

$$\alpha C \sigma^{*n} = \frac{1}{E} \frac{d\sigma}{dt} + C\sigma^n \tag{103}$$

where σ^* is the stress at $r = r^*$. The α is a function of time and it is generally difficult to solve equation (103). However, under two extreme conditions when $t = 0$ and for $t \to \infty$, the factor α is time independent with values α_e and α_c, respectively, as;

$$\alpha_e = \frac{\dot{\varepsilon}}{\dot{\varepsilon}^*} = \left(\frac{r^*}{r}\right)^{1/2} \qquad \text{for } t = 0 \tag{104}$$

$$\alpha_c = \frac{\dot{\varepsilon}}{\dot{\varepsilon}^*} = \left(\frac{r^*}{r}\right)^{\frac{n}{n+1}} \qquad \text{for } t \to \infty \qquad (105)$$

Equation (103) has been solved using these two expressions. The difference between the two cases is so small that the actual behaviour using the time dependent α can be predicted by using either constant value. As $\alpha \to \alpha_c$ the stress and strain rate fields become closer to the HRR type field characterised by equation (75). Using $\alpha = \alpha_c$ also predicts a larger time for stress redistribution, thus it will give a more conservative estimation of crack growth rates.

From the discussion above, the transition time t_w provides a conservative estimate for the stress around a crack tip to complete redistribution. If the time for a crack to start to grow is larger than t_w, the initial large stress and strain near the crack tip can be neglected, and it will be possible to use steady state creep conditions for the prediction of creep crack growth.

4. Experimental methods

The creep crack growth properties of materials are usually measured in tests which are carried out at constant load on fracture mechanics type specimens [79]. These specimens may contain an initial starter crack that has been introduced by fatigue or crack growth may take place from a sharp machined notch. The types of specimen that are used most often are compact tension (CT) (Figure 17), single edge notch tension (SENT), single edge notch bend (SENB), centre cracked plate (CCP) and double cantilever bend (DCB) test-pieces. Different thickness B and widths W are employed to represent different component dimensions. Frequently side-grooves (SG) are introduced to give a net thickness B_n to increase constraint and assist in promoting flat, straight fronted cracks. All these except the CT specimens have been used to a lesser extent [66-77,80-81]. Examples of crack growth against time curves that are commonly observed are shown in Figure 18.

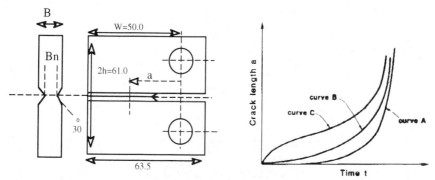

Figure 17: Typical dimensions of the CT specimen employed in the tests.

Figure 18: Examples of creep crack growth curves

These curves all indicate an increasing crack propagation rate with increase in time after crack growth has become well established. This is mainly because the magnitude of the stresses generated at the crack tip in these specimens increases with crack extension when they are subjected to a constant load.

The different trends depicted in the early stages of cracking can be attributed to a number of factors. Sometimes a prolonged incubation period (curve A) is observed prior to the onset of crack growth. During this stage damage gradually accumulates at the crack tip whilst crack blunting takes place. Eventually when sufficient damage has been introduced crack growth initiates. Curves B and C show the situation where crack growth starts immediately on loading. For curve B a progressively rising crack propagation rate is illustrated and for curve C an initial rapid cracking rate is observed which gradually reduces before adopting the common rising trend. The initial rapid cracking rate of curve C can be explained in two ways; by redistribution of stress at the crack tip from its initial elastic state to its steady state creep distribution and by primary creep. It is most likely that a shape like curve A will be exhibited by materials with appreciable creep ductility, and one like curve C by those with only limited ductility or by those which accumulate significant primary creep.

The extent of creep deformation will dictate whether the specimen will crack or fail by net-section rupture. Generally an increase in size and side-grooving will contain the deformation local to the crack tip thus enhancing plane strain and stable crack growth. The effects of temperature are more difficult to assess. Generally with increase in temperature there is an increase in the creep activation energy which could lead to either creep deformation and crack tip blunting or to creep crack growth or a combination of the two. The behaviour will be dictated by the extent of constraint at the crack tip. Figures 19-20 show an example of this behaviour for a DCB Aluminium alloy RR58 tested over the 100-200 °C range. Low temperatures and high temperatures reduce cracking rate and the peak of crack growth is achieved at 150 °C (figure 19). At temperatures greater than 150 °C slowing of crack growth is due to crack tip blunting whereas at lower temperatures the reduced cracking rate is due to creep embrittlement and low creep activation energy. The correlation versus C* shows a similar trend as shown in figure 20.

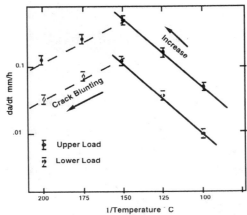

Figure 19: Effect of temperature on crack growth rate

There are two main methods of loading the specimen. In figure 21 a schematic example of the data that is obtained from the creep crack growth constant load testing is shown. Figure 21a shows an example of a compact tension specimen and the test variables that are collected for a static load test and figure 21b shows the same schematically but for a cyclic load test. In the latter case the value of displacement used is at the peak of each loading cycle. In figure

22 an example of data obtained from a constant displacement test is shown. In this case the load drops according to the extent of crack growth and /or creep deformation. The load line

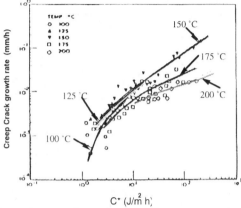

Figure 20: Dependence of crack growth rate on temperature correlated versus C* Aluminium Alloy RR58 tested between 100-200 °C using DCB specimens

displacement is measured in each case and the creep component of the total displacement measurement is used to calculate the experimental estimate of C* shown earlier. The standard tasting methods at high temperature are clearly presented in ASTM E1457 [79].

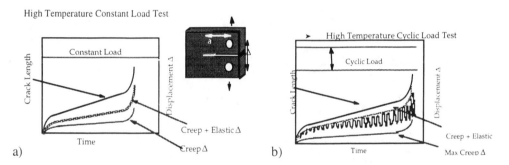

Figure 21: An example of experimental data obtained from Compact Tension creep crack growth tests under a) static and b) cyclic loading.

Typically crack length is measured optically if possible, and by means of the potential drop method either AC current or DC current. Usually AC is the preferred method but there are no set standards as regards their usage. Furthermore there are no established methods for the calibration of the potential drop versus crack length. Therefore a simple linear method is used to obtain crack length from potential drop readings during the test.

For crack length measurements the following method of linear data reduction is commonly used. The minimum number of crack length readings that are required are the initial and final crack lengths, which can be observed and measured directly from the cracked surface of the specimen (Figure 23). Additional intermediate points from either optical readings or specimen beach marking is preferred but not essential. When crack bowing is observed the average between the surface and the centre readings should be taken and the extent of bowing noted, this can be performed in accordance with ASTM E1457 [79].

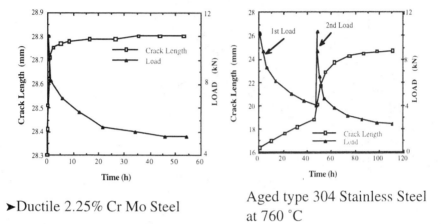

►Ductile 2.25% Cr Mo Steel

Aged type 304 Stainless Steel at 760 °C

Figure 22: Constant load displacement test showing crack growth and load relaxation .

Using these procedures crack length a is given by

$$a = \left(\frac{a_f - a_i}{V_f - V_i} \left(V - V_i \right) \right) + a_i \tag{106}$$

where a_i, a_f, V_i and V_f are respectively the initial and final crack length and potential drop readings.

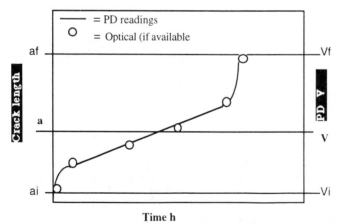

Figure 23: Linear calibration method for crack length.

Approximately 30-40 points are sufficient for each test with extra readings taken at the beginning and the end of the test where changes are more rapid. The corresponding time (in hours), crack length and LVDT measurements in mm are then saved as three columns. Each file should contain all the material property, specimen number and dimensions for further analysis. Analysis of the data is then carried out using the stress intensity factor K, the creep parameter C* and the reference stress, σ_{ref}. Methods of calculating these parameters have been shown in the earlier sections.

Metallurgical analysis is also performed to observe the damage and mode of cracking. An example of different cracking behaviour is shown in figures (24-26). In these figures the cracking behaviour is intergranular and therefore due to time dependent creep processes and with the exception of the extremely brittle 1/2CrMoV heat affected zone steel (figure 25) there is cracking ahead of the main crack tip suggesting local creep deformation and a ligament linkage mechanism for crack extension.

Figure 24: Crack growth micrograph for a static test of RR58 aluminium alloy tested at 150 °C showing intergranular creep crack with grain separation ahead of the main crack tip (1cm=30μm).

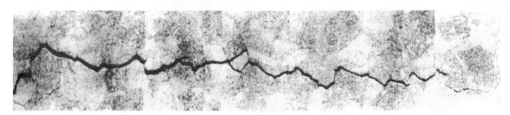

Figure 25: Crack growth micrograph for a static test of simulated heat affected zone 1/2CrMoV steel alloy tested at 565 °C showing a sharp intergranular creep crack (1cm=200μm).

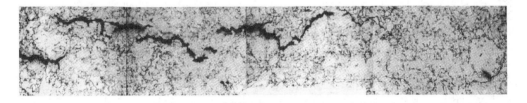

Figure 26: Crack growth micrograph for a static test of 21/4CrMo steel alloy tested at 535 °C showing a ductile intergranular creep crack with grain separation ahead of the main crack tip (1cm=150μm).

5. Modelling of crack initiation and growth at high temperatures

Methods have been presented for performing creep analyses of cracked and uncracked bodies. Fracture mechanics concepts have been developed for characterising stress distributions ahead of a crack in a component deforming by creep. In this section models are derived for describing the onset of cracking and crack propagation in terms of creep damage accumulation in a process zone ahead of the crack tip. Initially, models are developed for dealing with crack growth when a steady state distribution of damage has been generated at

the crack tip in materials which deform in secondary creep. In addition the importance of state of stress at the crack tip is examined. Afterwards, extensions to primary, secondary and tertiary creep are considered. Emphasis is then placed on the early stages of cracking. Criteria for the onset of cracking are introduced that are based on the attainment of 'a critical crack opening displacement or a specific amount of damage at the crack tip.

The object of this chapter is to present models of crack initiation and growth at elevated temperature using high temperature fracture mechanics. Both steady state and transient creep crack growth concepts will be employed. Based on these concepts, the determination of incubation times for creep crack growth is also shown. In addition, modelling of creep-fatigue crack growth under cyclic loading at elevated temperature is included.

5.1 Description of crack growth by fracture mechanics parameters

At elevated temperatures   where creep is. dominant, time-dependent crack growth is observed. The rate of this time-dependent crack growth is measurable and parameters are needed to predict it. In order to predict the crack behaviour in such materials, creep crack growth rate $\dot{a}$ (= da/dt) must be estimated using appropriate parameters. Several fracture parameters have been applied for this purpose. The most commonly used parameters are stress intensity factor K, the C^* integral and reference stress (net section stress) σ_{ref}. Crack growth rate $\dot{a}$ is usually given as follows using these parameters;

$$\dot{a} = AK^m \qquad\qquad (107)$$
$$\dot{a} = DC^{*\phi} \qquad\qquad (108)$$
$$\dot{a} = H\,\sigma_{ref}{}^p \qquad\qquad (109)$$

where A, D, H, m, ϕ and p are material constants which may depend on temperature and stress state. A suitable parameter to describe crack growth at elevated temperature will depend on material properties, loading condition, and time when crack growth is observed [80-84]. For example the formulation used in ASTM E1457 for a compact tension specimen is shown in table 2. Figure 27 represents typical relationships for creep crack in compact tension specimens growth versus a) K equation (107) for an aluminium alloy Al-2519 at 135 °C and b) C^* equation (108) for a 1CrMoV steel at 550 °C, showing an initial transient 'Tail' and a steady crack growth region.

If a material is elastic, immediately upon loading the stress distribution around the crack tip will be elastic. For this situation, the stress and strain at the crack tip are described by the stress intensity factor K. In a material that shows little creep deformation, the stress distribution will remain virtually unchanged by creep. Crack growth could also be characterised to a certain extent by the stress intensity factor K in such 'creep brittle' materials.

On the other hand, in a very 'creep ductile' material, where large creep strains can be dominant anywhere in the material, the singularity at the crack tip will be lost and correlation in terms of the reference stress on the uncracked ligament will be obtained. However, the reference stress is a parameter which describes the overall damage across an uncracked ligament and may characterise net section rupture of the ligament rather than creep crack growth.

For materials between these two extremes (creep brittle and very creep ductile), substantial creep deformation accompanies fracture, and stress redistribution will occur around a crack tip but a singularity by the crack will still remain. For this situation, creep crack growth will be characterised by $C*$. For engineering metals most experimental evidence suggests that the widest range of correlation is achieved with $C*$ [77, 81-88].

Table 2:   The formulae used for calculating K, reference stress, and the $C*$ parameter

Stress Intensity Factor for CT specimen (ASTM E1457)
$$K = \frac{P}{\sqrt{BB_N}\,W^{1/2}}\,\frac{2 + a/W}{(1 - a/W)^{3/2}}\,f(a/W)$$ where: $f(a/W) = 0.866 + 4.64(a/W) - 13.32(a/W)^2 + 14.72(a/W)^3 - 5.6(a/W)^4$
Reference stress for CT specimen
$$\sigma^{ref} = \frac{P}{mB_{eq}W}$$ where : $m = -(1 + \gamma\frac{a}{W}) + \sqrt{(1+\gamma)(\gamma(a/W)^2 + 1)}$ $$\gamma = \frac{2}{\sqrt{3}}$$ $$B_{eq} = B - \frac{(B - B_N)^2}{B}$$ B = thickness of the specimen B_N = the net thickness
$C*$ parameter (ASTM E1457)
$$C* = \frac{P\dot{V_c}}{B_N(W - a)}\frac{n}{n+1}(2 + 0.522\frac{W - a}{W})$$ where : $\dot{V_c}$ =creep displacement rate n = creep exponent

5.2 Steady state creep crack growth modelling

A model of creep crack growth under steady state conditions is presented in this section. The model has been proposed by Nikbin, Smith and Webster [77] and is known as the NSW model. The NSW model is based on stress and strain rate fields characterised by the $C*$ integral combined with a creep damage mechanism. For a material which deforms according to the Norton's creep law described by equation (72), the stress and strain rate distributions ahead of a crack are given by equations (75) and (76) respectively. When the creep damage accumulated within the creep zone shown in figure 28 reaches some failure value, the crack is postulated to progress. If material starts to experience creep damage when it enters the process zone at $r = r_c$ and accumulates creep strain ε_{ij} by the time it reaches a distance r from the crack tip, the condition for crack growth is given using the ductility exhaustion criterion as;

$$\varepsilon_{ij} = \int_{r=r_c}^{r} \dot{\varepsilon}_{ij} \, dt \tag{110}$$

Using the strain rate distribution by equation (87), the strain can be written as;

$$\varepsilon_{ij} = \int_{r=r_c}^{r} \varepsilon_0 \left[\frac{C^*}{I_n \sigma_o \dot{\varepsilon}_0 r} \right]^{\frac{n}{n+1}} \tilde{\varepsilon}_{ij} \frac{dt}{dr} \, dr \tag{111}$$

If a steady (constant) crack growth rate $\dot{a}_s$ is assumed,

$$\frac{dr}{dt} = -\dot{a}_s \tag{112}$$

Then equation (111) is analytically integrated as;

$$\varepsilon_{ij} = (n+1)\dot{\varepsilon}_0 \left[\frac{C^*}{I_n \sigma_o \dot{\varepsilon}_0} \right]^{\frac{n}{n+1}} \frac{\tilde{\varepsilon}_{ij}}{\dot{a}} \left[r_c^{\frac{1}{n+1}} - r^{\frac{1}{n+1}} \right] \tag{113}$$

When the creep ductility considering the stress state (multi-axial stress condition) is given by ε_f^*, substituting $\varepsilon_{ij} = \varepsilon_f^*$ at r = 0 into equation (112) gives;

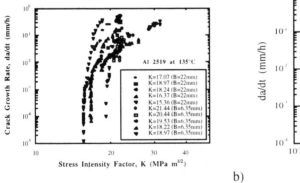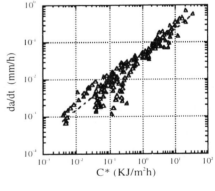

a)   b)

Figure 27: typical relationship for creep crack in compact tension specimens growth versus a) K and b) C* for an aluminium alloy Al-2519 at 135 °C and a 1CrMoV steel at 550 °C showing an initial transient 'Tail' and a steady crack growth region.

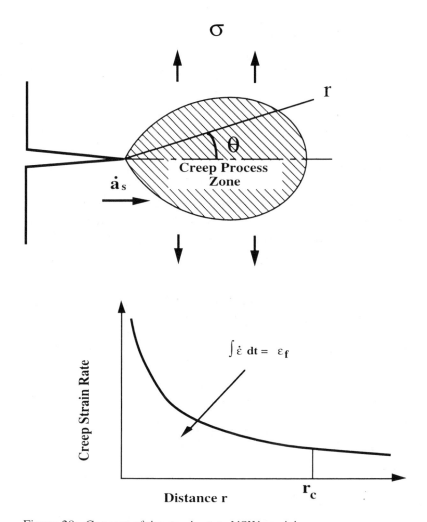

Figure 28: Concept of the steady state NSW model

$$\dot{a}_s = \frac{(n+1)\dot{\varepsilon}_0}{\varepsilon_f^*}\left[\frac{C^*}{I_n\sigma_o\dot{\varepsilon}_0}\right]^{\frac{n}{n+1}}\tilde{\varepsilon}_{ij}\,r_c^{\frac{1}{n+1}} \qquad (114)$$

The non dimensional function ε_{ij} is normalised so that its maximum equivalent becomes unity. Hence, assuming this maximum value, the constants D and ϕ in equation (108) become;

$$D = \frac{(n+1)\dot{\varepsilon}_0}{\varepsilon_f^*}\left[\frac{1}{I_n\sigma_0\dot{\varepsilon}_0}\right]^{\frac{n}{n+1}} r_c^{\frac{1}{n+1}}$$ (115)

$$\phi = \frac{n}{n+1}$$ (116)

The creep ductility considering the stress state $\varepsilon_f^{'}$ is equal to the uniaxial creep ductility ε_f in plane stress conditions. For plane strain conditions, the $\varepsilon_f^{'}$ could range between 25 to 80 times smaller than ε_f depending on the creep void growth model assumed [28,31] as described previously in section 2.5. The relationship between normalised ductility $\varepsilon_f/\varepsilon_f^{'}$ and the ratio of hydrostatic tensile stress over the equivalent stress ($\sigma_m/\overline{\sigma}$) for the two different models are shown in figure 29. It is apparent that an increase in hydrostatic tension causes a large reduction in creep ductility. The relationships are relatively insensitive to the creep index n. In the NSW model, $\varepsilon_f/\varepsilon_f^{'} = 50$ is recommended as a typical value for plane strain conditions [77]. This factor can be further refined to 25 for creep ductile materials [87].

Equation (108), which was developed in the above analysis, represents secondary creep only. In general, the cracking will exhibit a pseudo primary, secondary and tertiary creep behaviour. Material at the crack tip will also undergo three stages during its transverse of the creep zone similar to its uniaxial creep behaviour. In order to take all these stages into account, the analysis has been modified in the following way [76].

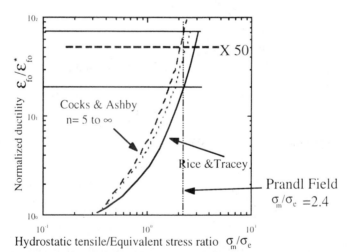

Figure 29: Influence of state of stress on creep failure strain and the power law creep equation (72) is also used as,

An average creep rate $\dot{\varepsilon}_A$ is defined as shown in figure 30 by;

$$\dot{\varepsilon}_A = \frac{\varepsilon_f}{t_r} \tag{117}$$

$$\dot{\varepsilon}_A = \dot{\varepsilon}_{0A}\left(\frac{\sigma}{\sigma_{0A}}\right)^n \tag{118}$$

where the material constants $\dot{\varepsilon}_{0A}$ and σ_{0A} are obtained from creep rupture data. Creep failure strain can be stress dependent, and if stress and rupture time are plotted as shown in figure 31 the rupture time will become,

$$t_r = \frac{\varepsilon_{f0}}{\dot{\varepsilon}_{0A}}\left(\frac{\sigma_{0A}}{\sigma}\right)^v \tag{119}$$

where ε_{f0} is the uniaxial creep rupture strain at stress σ_0. The creep rupture strain at stress σ is given from equations (117) to (119) as;

$$\varepsilon_f = \varepsilon_{f0}\left(\frac{\sigma}{\sigma_{0A}}\right)^{n-v} \tag{120}$$

When v is equal to n, the creep rupture strain is constant.

Nikbin et al. [76] used the time fraction rule and obtained the following equation to allow for the effects of primary and tertiary creep from equations (87) and (121) to (123). The time fraction rule and ductility exhaustion criteria give the same predictions when secondary creep dominates. Substitution of equation (117) and (118) into equation (114) gives

$$\dot{a}_s = \frac{(n+1)\dot{\varepsilon}_{0A}}{(n+1-v)\varepsilon^*_{f0}}\left[\frac{C^*}{I_n\sigma_{0A}\dot{\varepsilon}_{0A}}\right]^{v/(n+1)} r_c^{(n+1-v)/(n+1)} \tag{121}$$

where ε^*_{f0} is equivalent to creep ductility considering constraint effect.
For most steels, $n > 1$, usually 5 to 10, so that equation (121) is relatively insensitive to the value of r_c. Considering most engineering materials, equation (111) has been simplified [77]

as;
$$\dot{a} = \frac{3C^{*0.85}}{\varepsilon^*_f} \tag{122}$$

where $\dot{a}$ is in mm/hour, ε^*_f is strain as a fraction and C^* is in MJ/m²h.

5.3 Transient creep crack growth modelling

A transient creep crack growth model has also been proposed as described schematically in Figure 32 [76,77]. The differences between the steady crack growth model and the transient crack growth model are that transient creep crack growth starts from the undamaged state in the creep damage process zone, whereas the steady state model assumes a steady state distribution of damage in the process zone. Also creep damage accumulates in elements with width dr ahead of a crack tip. Steady state creep crack growth can be predicted analytically but numerical integration is necessary for the transient model.

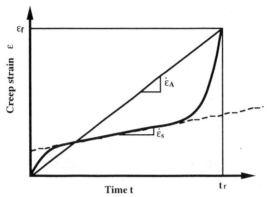

Figure 30: Simplification of the general creep curve

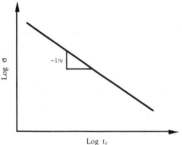

Figure 31 : Relationship between stress and rupture time

From the transient creep crack growth model, the initial crack growth rate $\dot{a}_0$ can be estimated by the following equation assuming secondary creep behaviour.

$$\dot{a}_0 = \frac{\dot{\varepsilon}_0}{\varepsilon_f^*}\left[\frac{C^*}{I_n\sigma_0\dot{\varepsilon}_0}\right]^{n/(n+1)} (dr)^{1/(n+1)} = \frac{1}{n+1}\dot{a}_s\left(\frac{dr}{r_c}\right)^{1/(n+1)} \qquad (123)$$

where dr is an increment of distance in the numerical integration for transient models and $\dot{a}_s$ is the steady state cracking rate. From the same reason described for the steady state model, the choice of dr appears to be unimportant for predicting creep crack growth rate because it is raised to a small fractional power

For element i, the model predicts a crack growth rate $\dot{a}_i$ of

$$\dot{a}_i = \frac{\dot{\varepsilon}_0}{\varepsilon_f^* - \varepsilon_{i,i-1}} \left[\frac{C^*}{I_n \sigma_0 \dot{\varepsilon}_0} \right]^{n/(n+1)} (dr)^{1/(n+1)} \qquad (124)$$

where $\varepsilon_{i,i-1}$ is the creep strain accumulated at $r = r_i = idr$ when the crack reaches $r = r_{i-1} = (i-1)dr$.

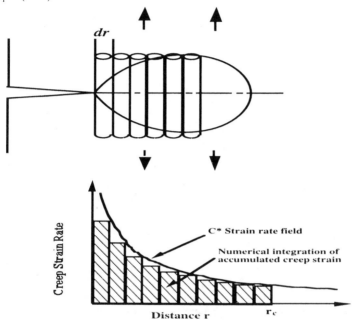

Figure 32: Concepts of the transient NSW model

Equations (123) and (124) can be further simplified [84,88] to relate the initial transient crack growth $\dot{a}_0$ to the steady state cracking rate $\dot{a}_s$ as

$$\dot{a}_o = \frac{1}{n+1} \left[\frac{dr}{r_c} \right]^{(1/n+1)} \dot{a}_s \qquad (125)$$

The ligament dr can be chosen to be a suitable fraction of r_c. However since dr/r_c is raised to a small power in equation (124) this will give

$$\dot{a}_o = \frac{1}{n+1} \dot{a}_s \qquad (126)$$

For most engineering materials therefore the initial crack growth rate is expected to be approximately an order of magnitude less than that predicted from the steady state analysis.

The cracking rate will progressively reach the steady state cracking rate as damage is accumulated.

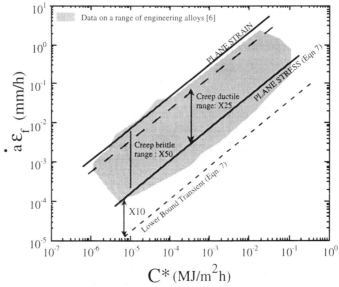

Figure 33: Material independent engineering showing a crack growth assessment diagram over the plane stress/strain and transient upper and lower bounds.

The regions of the crack growth behaviour seem to be covered adequately within the bounds of the present model. Figure 33, covering a whole range of materials [77,87] acts as a material independent engineering crack growth assessment diagram where cracking multiplied by the uniaxial ductility of the material. In addition, the transient region assuming that the initial cracking rate is given as $\dot{a}_0 \approx \dot{a}_s / 10$ can be introduced as a lower band to bound the initiation and the transition 'tail' region of damage development observed in many testpieces and components. The most conservative line to choose for life prediction would be the plane strain line. However for a narrow range of data sets using specific material, with fixed sizes and geometry an acceptable safety factor of 10 could possibly be used to predict cracking behaviour.

5.4 Consideration of creep damage in the uncracked ligament

For aged materials, creep damage may already have been accumulated before the creep process zone around a crack tip reaches it. Even in virgin materials, the ligament could be subject to significant creep damage if the crack growth is slow. A simple formula has been introduced to take this ligament damage into account as;

$$\dot{a} = \frac{D_0 C^{*\phi}}{1 - \omega} \qquad (127)$$

where D_0, ϕ are creep crack growth constants, and can be obtained from the NSW model. The term ω is the creep damage in the ligament which can be expressed when using the time fraction rule by,

$$\omega = \int \frac{dt}{t_r} \tag{128}$$

where t_r is the time to creep rupture of the ligament.

For a infinite centre crack plate (CCP), the following equations are obtained for crack growth into an undamaged ligament ($\omega=0$) as well as undamaged model ($\omega > 0$) [82,83].

$$\left\{ \left(\frac{a}{a_0} \right)^{1-\phi} - 1 \right\} = X \frac{t}{t_r} \quad \text{for the undamaged model}$$

$$\left\{ \left(\frac{a}{a_0} \right)^{1-\phi} - 1 \right\} = X \ln\left(\frac{1}{1-t/t_r} \right) \quad \text{for the damaged model} \tag{129}$$

where X is given for plane stress in a constant creep ductility material by, [83]

$$X = \left(\frac{\pi\sqrt{n}}{I_n} \right)^{n/(n+1)} \left(\frac{r_c}{a_0} \right)^{1/(n+1)} \tag{130}$$

5.5 Incubation time for the onset of crack growth

In creep crack growth experiments, an incubation period when the crack seems to be stationary can be often observed. Sometimes the incubation time occupies most of the life of a cracked body. Therefore, it may be important to incorporate the incubation time into creep crack growth predictions in practical applications. An incubation time prediction can be calculated from the NSW model [84,87-88]. Lower and upper bounds are obtained from the steady creep crack growth model and transient model as follows,

Lower bound: $$t_i(\text{Lower bound}) = \frac{r_c}{\dot{a}_s} = \frac{\varepsilon_f^*}{(n+1)\varepsilon_0} \left\{ \frac{I_n \sigma_0 \dot{\varepsilon}_0 r_c}{C*} \right\}^{n/(n+1)} \tag{131}$$

$$t_i(\text{Upper bound}) = \frac{r_c}{\dot{a}_0} = \frac{\varepsilon_f^*}{\varepsilon_0} \left\{ \frac{I_n \sigma_0 \dot{\varepsilon}_0 r_c}{C*} \right\}^{n/(n+1)} \left(\frac{r_c}{dr} \right)^{1/(n+1)}$$

Upper bound   (132)

$$= (n+1) \left(\frac{r_c}{dr} \right)^{1/(n+1)} t_i(\text{Lower bound})$$

Further the upper bound incubation can be obtained by taking $\dot{a}=\dot{a}_o$.

$$t_i = \frac{\varepsilon_{fo}^*}{\dot{\varepsilon}_o} \left[\frac{I_n \sigma_o \dot{\varepsilon}_o}{C*} \right]^{v/(n+1)} \tag{133}$$

Alternative estimates of an incubation period can also be obtained from the approximate creep crack growth rate equation (122). The prediction based on an approximate steady state cracking rate expression will be the lower bound of the incubation period.

$$t_i = \frac{r_c \varepsilon_f^*}{3C^{*0.85}}$$
(134)

When the initial crack growth rate is used instead of the steady state crack growth rate, the upper bound is determined as

$$t_i = \frac{(n+1)\, r_c \varepsilon_f^*}{3C^{*0.85}}$$
(135)

Figure 34 shows an example, for 1CrMoV steel tested at 550 °C, of how the model bounds the steady state and the transient crack growth rates and figure 35 for the incubation times using equation (134). The effects of geometry in figure 34 are shown as best fit lines but the overall inherent scatter (especially in the transient 'tail' region) in the data shows a much larger difference. The data lies in the plane stress region of cracking with the transient prediction indicating that the initial cracking rates could a factor of 10 slower than the steady state cracking rates.

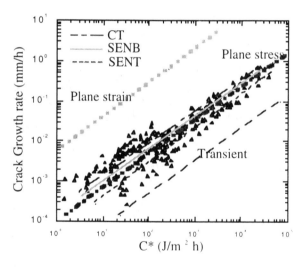

Figure 34: Predictions of crack growth rate versus C* for 1CrMoV steel tested at 550 °C using the steady state and the transient models.

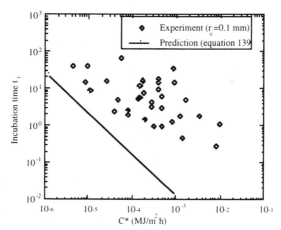

FIG. 35: Experimental and predicted incubation times for 1 CrMoV steel tested at 550 °C.

6. Creep-fatigue crack growth

Most engineering components which operate at elevated temperatures are subjected to non-steady loading during service. For example, electric power plant may be required to follow the demand for electricity and equipment used for making chemicals may undergo a sequence of operations during the production process. The power plants have to change their operating temperature and pressure to follow the demands of electricity need and to shut down and re-start for their routine maintenance as depicted in figure 36. Also, aircraft experience a variety of loading conditions during take-off, flight and landing. There may, in addition, be a superimposed high frequency vibration. Similarly, equipment that is subjected to predominantly steady operating conditions may experience transients during start-up and shut-down.

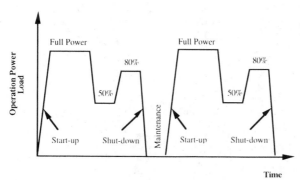

Figure 36: Example of operation scheme of a power plant

The gradual increase over the years in operating temperatures to achieve improved efficiency and performance from plant is causing materials to be used under increasingly arduous conditions. The first stage blades in aircraft gas-turbines can, for example, experience centrifugal stresses in the region of 150-200 MPa at gas temperatures that can exceed their incipient melting temperatures. Under these circumstances, survival is only possible if cooling is adopted to reduce average blade temperatures and coatings are used to limit erosion and environmental attack. The steep temperature gradients produced by the cooling will, however, introduce thermal stresses which will be regenerated each flight cycle and which can give rise to a mode of failure called thermal fatigue. The same situation can occur during rapid start-ups and shut-downs in thick sections of other components. Thermal fatigue is the type of failure that can occur by the repeated application of predominantly thermal stresses that are produced by the local constraint imposed by the surrounding material.

It is apparent, depending on the material of manufacture and the operating conditions, that creep, fatigue and environmental processes may contribute to failure. The dominating mode of failure in a particular circumstance will depend on such factors as material composition, heat-treatment, cyclic to mean load ratio, frequency, temperature and operating environment.

In the preceding sections, creep crack growth under steady loading was discussed. For the prediction of the crack growth in components at elevated temperatures, the interaction of creep and fatigue crack growth must be evaluated appropriately. In this section, the characteristics of fatigue crack growth are introduced first. Then a prediction of crack growth under the combination of creep and fatigue will be outlined.

6.1 Fatigue crack growth

Before considering creep-fatigue crack growth, some observations concerning the characteristics of fatigue crack growth will be discussed. References [89-96] give various description and views of the problems involved in fatigue and creep/fatigue conditions. Fatigue mechanisms usually dominate at room temperature. Procedures for measuring the fatigue crack propagation properties of materials at room temperature are described in reference [96]. Fatigue crack growth is usually observed as transgranular cracking at low temperature (as shown in figure 37 and is characterised by the stress intensity factor range ΔK using the Paris law [89]. At elevated temperature, transgranular cracks are also observed under relatively high frequency cycles(f>1Hz) and this fatigue crack growth rate can still be

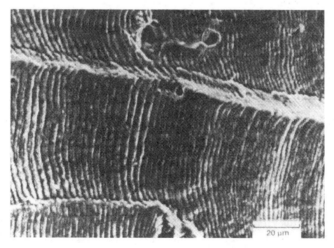

Figure 37: Example of striation and transgranular fatigue fracture surface of type 316 stainless steel.

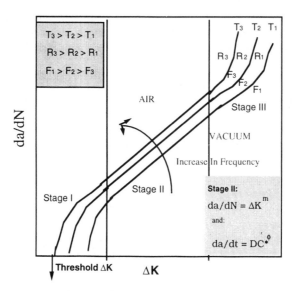

Figure 38: Cyclic crack growth at elevated temperatures

Transgranular cycle dependent and intergranular time dependent controlled cracking processes are identified. Conditions favouring each mechanism are clarified and it is shown that cumulative damage concepts can be applied to predict interaction effects. It is found that cyclic controlled processes are most likely to dominate at high frequencies and low R. Creep and environmental mechanisms, which are favoured at low frequencies, high temperatures and high R, are identified as contributing to the time dependent component of cracking. It is shown that these processes can enhance crack growth/cycle significantly and reduce component lives.

6.3 Creep-fatigue crack growth interaction

When alternating loads are applied to high temperature structures, the crack growth in the structures will be subject to both creep and fatigue. Interaction between creep and fatigue is expected under cyclic loading. Some of the causes of creep-fatigue interaction might be the enhancement of fatigue crack growth due to embrittlement of grain boundaries or weakening of the matrix in grains and enhancement of creep crack growth due to acceleration of precipitation or cavitation by cyclic loading [92,93]. The importance of creep-fatigue interaction effects is largely dependent on the material and loading conditions. Nevertheless, the simple linear summation rule for creep-fatigue crack growth defined by the following equation has been successfully applied to predict the crack growth in several engineering metals [97-99] is given as

$$\frac{da}{dN} = \left(\frac{da}{dN}\right)_{creep} + \left(\frac{da}{dN}\right)_{fatigue} = \frac{1}{3600f}\left(\frac{da}{dt}\right)_{creep} + \left(\frac{da}{dN}\right)_{fatigue} \qquad (139)$$

or

$$\frac{da}{dt} = \left(\frac{da}{dt}\right)_{creep} + \left(\frac{da}{dt}\right)_{fatigue} = \left(\frac{da}{dt}\right)_{creep} + 3600 f \left(\frac{da}{dN}\right)_{fatigue} \qquad (140)$$

where da/dN is crack growth per cycle in mm/cycle, da/dt is crack growth rate in mm/hour, and f is frequency in Hz.

Effects of the influence of frequency [97-99] on crack growth/cycle in a nickel base alloy (AP1) at 700°C are shown in figures 39 and 40. Figure 39 indicates a dependence of da/dN on frequency at R = 0.7. In the steady state cracking region crack growth can be described by the Paris law (equation (137)) with m ≈ 2.5. This value is within the range expected for room temperature behaviour. Correlation of results obtained at 20 MPa √m for specimen thickness (B=25mm) is depicted in figure 40. Furthermore in figure 40 the two lines show the relative behaviour of the creep and the fatigue components of da/dN.

The slope of -1 indicates time dependent creep cracking and the horizontal line indicates fatigue control of da/dN. Therefore by adding the two components is clear that at high frequency creep is seen to have third order effect on cracking rate and conversely at low frequencies fatigue has a third order effect. From metallurgical and fractographic investigations performed on the alloy tested in the creep and creep fatigue range similar qualitative conclusions can be reached with respect to the mode of the creep fatigue interaction. Figure 41 show the fractographs for the nickel-base superalloy AP1 tested at 700 °C. There is a transition from intergranular cracking at f=0.001 Hz to transgranular cracking at 10 Hz.

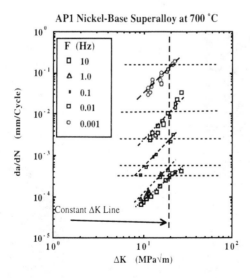

Figure 39: Frequency dependence of fatigue cracking at high temperatures for AP1 nickel-base superally tested at 700 °C and R=0.7

➤ da/daN = (da/dN) + (da/dt)/f
   Total   =   Fatigue + Creep

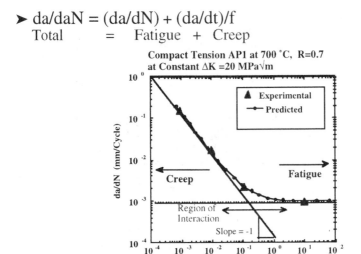

Figure 40: Fatigue crack growth sensitivity to frequency in an AP1 nickel-base superalloy tested at 700 °C.

The intermediate frequencies show a mixture of intergranular and transgranular cracking modes. These suggest that the two mechanism work in parallel and that cumulative damage concepts proposed above can well describe the total cracking behaviour.

➤ Surface Micrographs in Creep/Fatigue Interaction

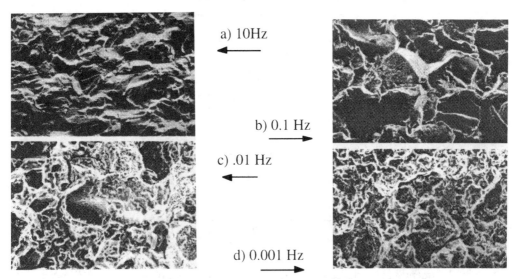

Transgranular (Fatigue) >> Intergranular (Creep)

Figure 41: Effects of frequency on mode of failure for AP1 astroloy Nickel-base superalloy tested at 700 °C.

The effects of R-ratio on creep/fatigue interaction is shown in figures 42 and 43. Generally as shown in figure 42 an increase in the R-ratio reduces the ΔK needed for crack growth per cycle and increase in frequency reduced the da/dN. Figures 43a,b compare the da/dN in terms of frequency in terms of constant ΔK and constant K_{max}. Figure 43a shows a dependence of da/dN versus frequency on R-ratio. It is clear that at low frequencies cracking is independent of R-ratio when plotted in terms of K_{max} as shown in figure 43b. This suggests that creep failure dominates at maximum load for low frequencies.

In both figures 40 and 43 it is clear that the interaction region of creep fatigue is contained within a small frequency band of 1 decade in the region of 0.1 to 1 Hz. From experimental and data analysis performed [97-99] it has been shown that static crack growth data correlates well with low frequency (≤ 0.001 Hz) data. Therefore equations (143-144) can be used to incorporate static data with cyclic data. This method is useful for cases where no cyclic data exists and only static and high frequency data is available.

The effects of temperature and the predictions of creep/fatigue using static data are shown in figure 44. This figure shows for a range of test temperatures both static and cyclic data for a martensitic alloy FV448. Whilst the high frequency data varies very little with temperature and frequency the effects of predicted static and low frequency data shift to lower frequencies indicating that fatigue will dominate failure at low frequencies.

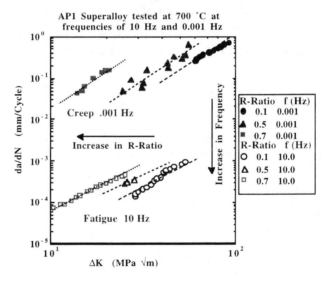

Figure 42: Dependence of crack growth on frequency and R-Ratio for an AP1 nickel-base superalloy tested at 700 °C.

The interpretations of elevated temperature cyclic crack growth behaviour have so far been presented mainly in terms of linear elastic-fracture mechanics concepts. It is anticipated that this approach is adequate when fatigue and environmental processes dominate as stress redistribution will not take place at the crack tip. When creep mechanisms control, stress redistribution will occur in the vicinity of the crack tip and use of the creep fracture mechanics parameter C* becomes more appropriate.

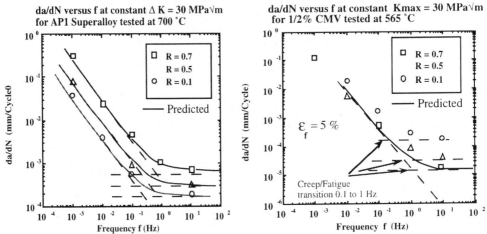

Figure 43: Fatigue crack growth sensitivity to frequency and R-ratio in an AP1 nickel-base superalloy tested at 700 °C at a) constant ΔK=30MPa√m and b) constant Kmax= 30Mpa√m.

In the region where the Paris law is relevant, substitution of the C* relation to crack growth into equation (138) allows cyclic crack growth under creep-fatigue loading conditions to be established from

$$da/dN = C\Delta K^m + DC*^\phi / f \qquad (141)$$

Alternatively, if the approximate expression in equation (122) is employed for the creep component of cracking, crack growth/cycle becomes,

$$da/dN = C\Delta K^m + 3C*^{0.85} / \varepsilon_f^* .3600 f \qquad (142)$$

and for static loading

$$da/dN = C\Delta K^m + 3C*^{0.85} / \varepsilon_f^* .7200 f \qquad (143)$$

where da/dN is in mm/cycle with frequency in Hz. To allow for crack closure effects, ΔK in these equations can be replaced by ΔK_{eff}. Similarly ΔJ can be used instead of ΔK when plastic deformation is significant. In order to make predictions of creep-fatigue crack growth in components it is necessary to be able to calculate ΔK and C* as crack advance occurs. The same procedures that are employed for estimating K and C* under static loading can be employed.

The effects of constraint due to geometry and specimen size can be described using the cumulative method and over a range of frequencies or using C* at low frequencies where creep dominates. Figure 45 shows the effects of geometry and specimen size on crack growth at constant ΔK over a range of frequencies for the AP1 alloy at 700 °C.

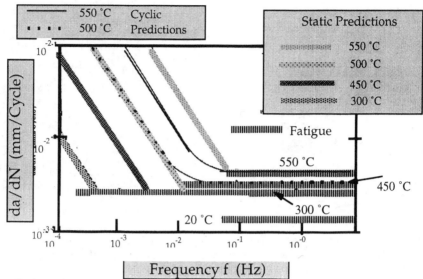

Figure 44: Prediction of da/dN for a Martensitic steel from static data, over a temperature range of 20-550 °C using static and cyclic data at a constant $\Delta K=50MPa\sqrt{m}$.

At high frequencies there are no size and geometry effects however at low frequencies the Corner Crack Tension (CCT) specimens, shown in figure 46a exhibit lower *da/dN* compared to the thin CT specimens and this in turn shows a lower *da/dN* compared to the 25mm thick CT. Therefore constraint effects are exaggerated under creep conditions and increase in constraint tends to increase cracking rates.

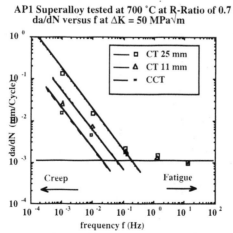

Figure 45: Geometry and size effects in the creep/fatigue crack growth of AP1 nickel-base superalloy tested at 700 °C.

Corner Cracked Tension CCT

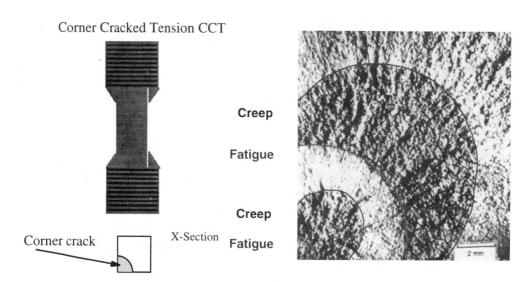

Figure 46: a) Corner Crack Tension specimen AP1 Superalloy tested at 700 °C b) and a fractograph showing influence of creep and fatigue on the shape and texture of the crack front.

The effects of geometry on crack growth for the static and low frequency data are plotted against C* in figure 47. It is clear that cracking rate is state of stress controlled in the creep range. Figure 46b shows a fractograph exhibiting the effects of creep and fatigue on the actual cracking of the CCT. In this case the frequency was varied between 10Hz and 0.001 Hz. Where fatigue dominates the crack front profile is approximately a quadrant and the cracking rate is the same in the inner section and the surface. When frequency is reduced the crack leads in the centre and where the surface is more in plane stress crack growth rate reduces. The effect is reversed when the frequency is increased once again.

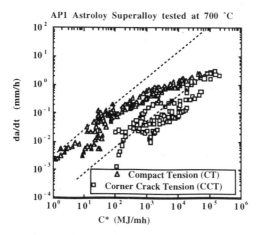

Figure 47: Effect of geometry in creep crack growth of AP1 Nickel-base superalloy tested at 700 °C.

Finally it has been shown that high-temperature crack growth occurs by either cyclic-controlled or time-dependent processes. Over the limited range where both mechanisms are significant, a simple cumulative damage law can be employed to predict behaviour. Interpretations have been developed in terms of linear elastic and non linear fracture mechanics concepts. Linear elastic fracture mechanics descriptions are expected to be adequate when fatigue and environmental processes dominate. When creep mechanisms control stress redistribution takes place in the vicinity of the crack tip and use of the creep fracture mechanics parameter C* should be employed for characterising the creep component of cracking.

7 Relevance to life assessment methodology

In this section the relevance of the what has been discussed above is highlighted by describing the problems associated with creep and creep/fatigue in plant components. For new plant design procedures are required to avoid excessive creep deformation and fracture in electric power generation equipment, aircraft gas turbine engines, chemical process plant, supersonic transport and space vehicle applications. There is also a continual trend towards increasing the utilisation and efficiency of plant for economic reasons. This can be achieved in a number of ways. One is to fabricate components out of new materials with improved creep properties. Another is to increase the temperature of operation which may itself require the introduction of new materials.

By extending plant life and to avoid premature retirement on the basis of reaching the design life which may have been obtained previously using conservative procedures is an economic reality for most industrial operators. Many electric power generation and chemical process facilities are designed to last for 30 years or more assuming specific operating conditions. It is possible that if they are used under less severe conditions to those anticipated or a re-evaluation using the modern understanding of creep and fracture is performed it may justify a further extension of the plant life.

To avoid excessive creep deformation and prevent fracture are the two important high temperature design considerations. In most industries these objectives are achieved by the application of design codes which incorporate procedures for specifying maximum acceptable operating stresses and temperatures. The maximum values allowed depend upon the type of component and consequences of failure. The magnitudes chosen for the safety factors are determined by experience and are dependent on the type of calculation performed and whether average or minimum properties data are employed. Different safety factors may also be applied to normal operating conditions, frequent, infrequent or emergency excursions.

Most high temperature design codes have been developed from those that have been produced for room temperature applications. They are therefore aimed at avoiding failure by plastic collapse, fatigue and fast fracture as well as creep and it is possible to define temperatures, which vary somewhat between codes because of the different procedures employed, below which creep need not be considered for particular classes of materials. Guidance has been in existence for some while to guard against fast fracture and fatigue below the creep range and is now becoming available for elevated temperature situations where creep is of concern.

The high temperature design codes in current use have been developed principally for application to defect free equipment. There is an increasing trend for critical components to be subjected to non-destructive examination to search for any possible flaws. Flaws can be detected by visual, liquid penetrant, magnetic particle, eddy current, electrical potential, radiographic and ultrasonic means depending on whether they are surface breaking or buried. Consequently there is a requirement for establishing tolerable defect sizes. Also the improving sensitivity of these techniques is causing smaller and smaller flaws to be found and the question of whether they must be removed, repaired or can be left is being encountered more frequently.

Finally during operation different equipment can experience a wide range of types of loading. For example, stress and temperature can be cycled in-phase or out-of-phase, a stress can be applied or removed at different rates and dwell periods can be introduced during which stress or strain are maintained constant. In these circumstances, it is important that the processes governing failure are properly understood so that the appropriate predictive models and equations can be employed.

8. References

1. Webster, G. A., And Ainsworth, R. A., "high temperature component life assessment", Chapman & Hall, 1994.
2. Webster, G.A., 'A widely applicable dislocation model for creep', Phil. Mag., 14, 1966, 775-783.
3. Dorn, J.E., Ed, 'Mechanical behaviour of materials at elevated temperatures', McGraw-Hill, Inc., New York, 1961.
4. Kennedy, A.J., 'Processes of creep and fatigue in metals', Wiley, New York, 1962.
5. Garofalo, F., 'Fundamentals of creep and creep-rupture in metals', MacMillan, New York, 1965.
6. Mclintock, F.A. and Argon, A.S., 'Mechanical behaviour of materials', Addison-Wesley, Massachusetts, 1966.
7. Gemmill, M.G., The technology and properties of ferrous alloys for high temperature use', Newnes, London 1966
8. Gittus, J., 'Creep, viscoelasticity and creep fracture in solids', Applied Science, London, 1975.
9. Frost, H.J., and Ashby, M.F., 'Deformation-mechanism maps', Pergamon Press, Oxford, 1982.
10. Riedel, H., 'Fracture at high temperatures', Springer-Verlag, Berlin, 1987.
11. Cadek, J,'Creep in Metallic Materials', Elsevier, Amsterdam, 1988.
12. Johnson, W.G. and Gilman, J.J., 'Dislocation velocities, dislocation densities and plastic flow in lithium fluoride crystals', J. App Phys, 30, no. 2, 1959, 129-144.
13. Haasen, P., 'III Dislocation mobility and generation, dislocation motion and plastic yield of crystals', Discussions of the Faraday Soc, no. 38, 1964, 191-200.
14. Webster, G.A., Cox, A.P.D. and Dorn, J.E., 'A relationship between transient and steady-state creep at elevated temperatures', Met Sci J, 3, 1969, 221-225.
15. Evans, R.W. and Wilshire, B., 'Creep of metals and alloys', Inst. of Metals, London, 1985.
16. Ashby, M.F., Gandhi, C. and Taplin, D.M.R., 'Fracture-mechanism maps and their construction for FCC metals and alloys', Acta Met, 27, 1979, 699-729.
17. Gandhi, C and Ashby, M.F., 'Fracture-mechanism maps for materials which cleave : FCC, BCC and HCP metals and ceramics', ib id, 1565-1602.

18. Gittus, J., 'Cavities and cracks in creep and fracture', Applied Science, London, 1981.
19. Monkman, F.C. and Grant, N.J., 'An empirical relationship between rupture life and minimum creep rate in creep-rupture tests', Proc Am Soc Testing Materials, 56, 1956, 593-620.
20. Finnie, I. and Heller, W.R., 'Creep of engineering materials', McGraw-Hill, New York, 1959.
21. Lubahn, J.D. and Felgar, R.P., 'Plasticity and creep of metals', Wiley, New York, 1961.
22. Johnson, A.E., Henderson, J. and Khan, B., 'Complex stress creep, relaxation and fracture of metallic alloys', HMSO, London, 1962
23. Odqvist, F.K.G., 'Mathematical theory of creep and creep rupture', Oxford University Press, Oxford, 1966.
24. Rabotnov, Yu N., 'Creep problems in structural members', (Ed. F.A. Leckie), North Holland, Amsterdam, 1969.
25. Penny, R.K. and Marriott, D.C., 'Design for creep', McGraw-Hill, London, 1971.
26. Boyle, J.T. and Spence, J., 'Stress analysis for creep', Butterworths, London, 1983.
27. Viswanathan, R., 'Damage mechanisms and life assessment of high-temperature components', ASM International, Metals Park, Ohio, 1989.
28. Cocks, A.C.F. and Ashby, M.F., 'Intergranular fracture during power-law creep under multiaxial stress', Met. Sci., 14, 1980, 395-402.
29. Smith, D.J. and Webster, G.A., 'Fracture mechanics interpretations of multiple-creep cracking using damage-mechanics concepts', Mat. Sci and Tech., 1, 1985, 366-372.
30. Kachanov, L.M., 'Introduction to continuum damage mechanics', Kluwer Academic Publishers, Dordrecht, 1986.
31. Rice, J.R. and Tracey, D.M., 'On the ductile enlargement of voids in triaxial stress fields', J. Mech. Phys. Solids, 17, 1969, 201-217.
32. Forrest, P.J., 'Fatigue of metals', Addison-Wesley, Reading, USA, 1962.
33. Forsyth, P.J.E., 'The physical basis of metal fatigue', Elsevier, New York, 1969.
34. Fuchs, H.O. and Stephens, R.I., 'Metal fatigue in engineering', J. Wiley, New York, 1980.
35. Hertzberg, R.W., 'Deformation and fracture mechanics of engineering materials', J. Wiley, New York, 1983.
36. Bressers, J. (Ed) 'Creep and fatigue in high temperature alloys', Applied Science, Barking, UK, 1981.
37. Coffin, L.F., 'Fatigue at high temperature' in 'Fatigue at elevated temperatures', ASTM STP 520, 1973, 5-34.
38. Manson, S.S., 'A challenge to unify treatment of high temperature fatigue - a partisan proposal based on strainrange partitioning in fatigue at elevated temperatures, in Fatigue at elevated temperatures', ASTM STP 520, 1973, 744-775.
39. Irwin, Fracture dynamics, in Fracturing of metals, 1948, ASM: p. 147-166.
40. Rice, J. R. and Rosengren, G. F., Plane strain deformation near a crack tip in a power-law hardening material. Journal of Mechanics and Physics of Solids, 1968. 16: p. 1.
41. Rice, J. R., A Path Independent Integral and the Approximate Analysis of Strain Concentration by Notches and Cracks. Journal of Applied Mechanics, 1968. E35(ASME): p. 379-386.
42. Williams, M. L., On the stress distribution at the base of a stationary crack. Journal of Applied Mechanics, 1957. 24: p. 109-114.
43. Sih, G. C., Handbook of Stress Intensity Factors. 1973, Bethlehem, Penn: Institute of Fracture and Solid Mechanics, Lehigh University.

44. Rooke, D. P. and Cartwright, D. J., Compendium of Stress Intensity Factors. 1976, London: Her Majesty's Stationary Office.
45. Tada, H., Paris, P. C., and Irwin, G. R., The Stress Analysis of Cracks Handbook. Second Edition ed. 1985, Hellertown, Pa.: Del Research.
46. Hutchinson, J. W., Singular behaviour at the end of a tensile crack in a hardening material. Journal of the Mechanical and Physics of Solids, 1968. 16: p. 13-31.
47. Griffith, A. A., The Phenomena of Rupture and Flow in Solids. Philosophical Transactions of Royal Society of London, 1920. A-221(163-198).
48. McClintock, F. A., Mechanics in Alloy Design, in Fundamental Aspects of Structural Alloy Design, R. I. Jaffee and B. A. Wilcox ed. 1977, Plenum Press: New York.
49. Parks, D. M., The Virtual Crack Extension Method for Non-linear Material Behaviour. Computer Methods in Applied Mechanics and Engineering, 1977. 12: p. 353-364.
50. Hellen, T. K., International Journal of Numerical Mathematics and Engineering, 1975. 9: p. 187.
51. He, M. Y. and Hutchinson, J. W., The penny-shaped crack and the plane strain crack in an infinite body of power-law material. Journal of Applied Mechanics, 1981. 48: p. 830-840.
52. Kumar, V., German, M. D., and Shih, C. F., An Engineering Approach for Elastic-Plastic Fracture Analysis, NP-1931, 1981,Electric Power Research Institute
53. Kumar, V. and Shih, C. F., Fully Plastic Crack Solutions, Estimation Scheme, and Stability Analyses for the Compact Specimen. ASTM STP, 1980. 700(American Society for Testing and Materials): p. 406-438.
54. Kumar, V., et al., Estimation Technique for the Prediction of Elastic-Plastic Fracture of Structural Components of Nuclear Systems, RP-1237-1, 1982,GE/EPRI
55. Kumar, V. and German, M. D., Elastic-Plastic Fracture Analysis of Through-Wall and Surface Flaws in Cylinders, NP-5596, 1988,GE/EPRI
56. Shih, C. F. and Needleman, A., Fully Plastic Crack Problems. Journal of Applied Mechanics, 1984. 51: p. 48-56.
57. Shih, C. F. and Needleman, A., Fully Plastic Crack Problems -Part 2: Application of Consistency Checks-. Journal of Applied Mechanics, Transaction of ASME, 1984. 51: p. 57-64.
58. Kumar, V., et al., Advances in Elastic-Plastic Fracture Analysis, NP-3607, 1984,GE/EPRI
59. Yagawa, G., Takahashi, Y., and Ueda, H., Three-dimensional fully plastic solutions for plates and cylinders with through-wall cracks. Journal of Applied Mechanics, 1985. 52: p. 319-325.
60. Ainsworth, R. A. and Goodall, I. W., Defect Assessment at Elevated Temperature. Journal of Pressure Vessel Technology, 1983. 105: p. 263-268.
61. Ainsworth, R. A., Some Observations on Creep Crack Growth. International Journal of Fracture, 1982. 18: p. 147-159.
62. Ainsworth, R. A., The Assessment of Defects in Structures of Strain Hardening Material. Engineering Fracture Mechanics, 1984. 19(4): p. 633-642.
63. Goodall, I. W., et al., Development of High Temperature Design Methods based on Reference Stress. Journal of Engineering Materials and Technology, 1979. 101: p. 349-355.
64. Miller, A. G., Review of Limit Loads of Structures Containing Defects. International Journal of Pressure Vessel and Piping, 1988. 32: p. 197-327.
65. ASTM, Standard Test for JIC, a Measure of Fracture Toughness, ASTM E813, 1987.
66. Nikbin, K. M., Webster, G. A., and Turner, C. E. A Comparison of Methods of Correlating Creep Crack Growth. in ICF4. 1977. Waterloo, Canada:

67. Rice, J. R., Paris, P. C., and Merkle, J. G., Some Further Results of J-integral and Analysis Estimates, in Progress in Flaw Growth and Fracture Toughness Testing, STP 536, 1973, ASTM: p. 231.
68. Smith, D. J. and Webster, G. A., Estimates of the C* parameter for crack growth in creeping materials. ASME STP, 1983. 803: p. I-654-I-674.
69. Riedel, H. and Rice, J. R., Tensile Cracks in Creeping Solids, in Fracture Mechanics, P. C. Paris ed. STP 700, 1980, ASTM: p. 112-130.
70. Riedel, H., Fracture at High Temperature. 1st ed. Material Research and Engineering, eds. B. Ilschner and N. J. Grant. 1986, Berlin: Springer-Verlag Berlin. 418.
71. Ehlers, R. and Riedel, H. in Advances in Fracture Research, Proceeding of ICF5. 1981. Pergamon Press.
72. Ainsworth, R. A. and Budden, P. J., Crack tip fields under non-steady creep conditions - I. Estimates of the amplitudes of the fields. Fatigue and Fracture of Engineering Materials and Structures, 1990. 13: p. 263-276.
73. Saxena, A. Crack Growth under Non Steady-state Conditions. in 17th ASTM National Symposium on Fracture Mechanics. 1984. Albany, NY
74. Bassani, J. L., Donald, D. E., and Saxena, A., Evaluation of the Ct Parameter for Characterizing Creep Crack Growth Rate in the Transient Regime. ASTM STP, 1989. 995: p. 7-26.
75. Webster, G. A., et al., High Temperature Component Life Assessment. 1991, London: Imperial College of Science, Technology and Medicine - Lecture Notes.
76. Nikbin, K. M., Smith, D. J., and Webster, G. A., Prediction of Creep Crack Growth From Uniaxial Creep Data. Proc. of Royal Society, London, 1984. A396: p. 183-197.
77. Nikbin, K. M., Smith, D. J., and Webster, G. A., An Engineering Approach to the Prediction of Creep Crack Growth. Journal of Engineering Materials and Technology, 1986. 108: p. 186-191.
78. Nuclear Electric, Assessment Procedure for the High Temperature Response of Structures, R-5, 1990.
79. ASTM, Standard Test Method for Measurement of Creep Crack Growth Rates in Metals, ASTM E1457-92, 1992.
80. Ellison, E. G. and Harper, M. P., Creep Behaviour of Components Containing Cracks - a Critical Review. Journal of Strain Analysis, 1978. 13: p. 35-51.
81. Nikbin, K. M., Smith, D. J., and Webster, G. A., Influence of Creep Ductility and State of Stress on Creep Crack Growth, in Advances in life prediction methods at elevated temperatures, D. A. Woodford and J. R. Whitehead ed. 1983, ASME: New York. p. 249-258.
82. Nishida, K., Nikbin, K. M., and Webster, G. A., Influence of Net Section Damage on Creep Crack Growth. Journal of Strain Analysis, 1989. 24(2): p. 75-82.
83. Nishida, K. and Webster, G. A., Interaction between build up of local and remote damage on creep crack growth, in Creep and Fracture of Engineering Materials and Structures, B. Wilshire and R. W. Evans ed. 1990, Inst. Metals: Swansea. p. 703-713.
84. Austin, T. S. P. and Webster, G. A., Prediction of Creep Crack Growth Incubation Periods. Fatigue and Fracture of Engineering Materials and Structures, 1992. 15(11) p. 1081-1090.
85. Ohji, K., Ogura, K., and Kubo, S., Stress field and modified J-integral near a crack tip under conditions of confined creep deformation. Zairyo (in Japanese), 1980. 29 (320) p. 467-471.

86.   Saxena, A. Evaluation of C* for the Characterization of Creep-Crack Growth Behavior in 304 Stainless Steel. in Twelfth Conference. 1980. American Society for Testing and Materials.

87.   Nikbin, K. M., "Consideration of Safety Factors in the Life Extension Modeling of Components Operating at High Temperatures"Effects of Product Quality Control and Design Criteria on Structural Integrity, ASTM STP 1337, R. C. Rice, D. E. Tritsch, Eds., American Society for Testing and Materials, 1998.

88.   Nikbin, K. M. Transition Effects in Creep-Brittle Materials. in Mechanics of Creep Brittle Materials-2. 1991. Leicester, England

89.   Paris, P. C., Fracture Mechanics in the Elastic Plastic Regime. ASTM STP, 1977. 631(American society for Testing and Materials): p. 3-27.

90 . Forman, R. G., Kearney, V. E., and Engle, R. M., Numerical analysis of Crack Propagation in a Cyclic-loaded Structures. ASME Transaction, Journal of Basic Engineering, 1967. 89(D): p. 459.

91.   Paris, P. C., Gomez, M. P., and Anderson, W. E., A Rational Analytic Theory of Fatigue. Trend in Engineering, 1961. 13: p. 9-14.

92.   Kaneko, H., et al. Study on Fracture Mechanism and a Life Estimation Method for Low Cycle Creep-Fatigue Fracture of Type 316 Stainless Steels. in Low cycle fatigue and elasto-plastic behaviour of materials -3. 1992. Berlin, FRG: Elsevier Applied Science.

93.   Nakazawa, T., et al. Study on Metallography of Low Cycle Creep Fatigue Fracture of Type 316 Stainless Steels. in Low cycle fatigue and elasto-plastic behaviour of materials -3. 1992. Berlin, FRG: Elsevier Applied Science.

94.   Nikbin, K. M. and Webster, G. A., Prediction of Crack Growth under Creep-Fatigue Loading Conditions, STP 942, 1987, American society for Testing and Materials: Philadelphia. p. 281-292.

95.   Austin, T. S. P. and Webster, G. A., Application of a Creep-Fatigue Crack Growth Model to Type 316 Stainless Steel, in Behaviour of Defects at High Temperatures, R. A. Ainsworth and R. P. Skelton ed. 15, 1993, Mechanical Engineering Publications Limited: London. p. 219-237.

96.   ASTM E647-86a, 'Standard test method for measuring fatigue crack growth rates', Book of Standards, Am.Soc.Testing and Matls., Philadelphia, 1987, 03.01, 899-926.

97.   Nikbin, K.M. and Webster, G.A., 'Creep-fatigue crack growth in a nickel base superalloy' in 'Creep and fracture of engineering materials and structures', (Eds B. Wilshire and D.R.J. Owen). Pineridge Press, Swansea, 1984, 1091-1103.

98.   Winstone, M.R., Nikbin, K.M. and Webster, G.A., 'Modes of failure under creep/fatigue loading of a nickel-base superalloy', J. Matls Sci, 1985, 20, 2471-2476.

99.   Dimopulos, V., Nikbin, K.M. and Webster, G.A., 'Influence of cyclic to mean load ratio on creep/fatigue crack growth', Met. Trans. A, 1988, 19A, 873-880.